Mysteries of the Universe

Great Mysteries

Book Club Associates
London

Mysteries of the Universe
by Roy Stemman

Series Coordinator   John Mason
Design Consultant   Günther Radtke
Art Director   Peter Cook
Editors   Sally Burningham
     Damian Grint
Editorial Consultant   Beppie Harrison
Picture Reasearch   Marian Pullen, Frances Vargo

This edition published 1980 by
Book Club Associates
By arrangement with Aldus Books Limited
First published in the United Kingdom
in 1978 by Aldus Books Limited
17 Conway Street, London W1P 6BS

Printed and bound in Hong Kong
by Leefung-Asco Printers Limited

Introduction

Astronomy is the oldest of the sciences. It is also the most modern. Yet the amount we actually know about the vast universe in which we live is amazingly small. Men have walked on the surface of the Moon: satellites have probed Mars and Venus. But the mysteries that surround us in the inky blackness of space are still as vast as space itself. What, for example, were the signals from outer space that radio astronomers received so regularly in the 1960s? What exactly is that most awe-inspiring object in the entire universe—the so-called "black hole"? Has the universe always existed: will it continue for ever? Just how much do we know about our nearest neighbors the planets, which like Earth, are in orbit around our sun? This book gives up-to-date answers where they are available, and examines the theories where there is only conjecture. The author then tackles the greatest of all space mysteries—are we alone in the universe? Beginning with the first well-documented sightings of flying saucers in the years directly after World War II, he covers the whole range of evidence and theory, from Erich von Däniken to the Condon Report. Finally, he describes how scientists are now beaming messages telling of our existence across the stars in the gigantic Milky Way. If there is an answer, how might it come. More importantly, what might it be?

Contents

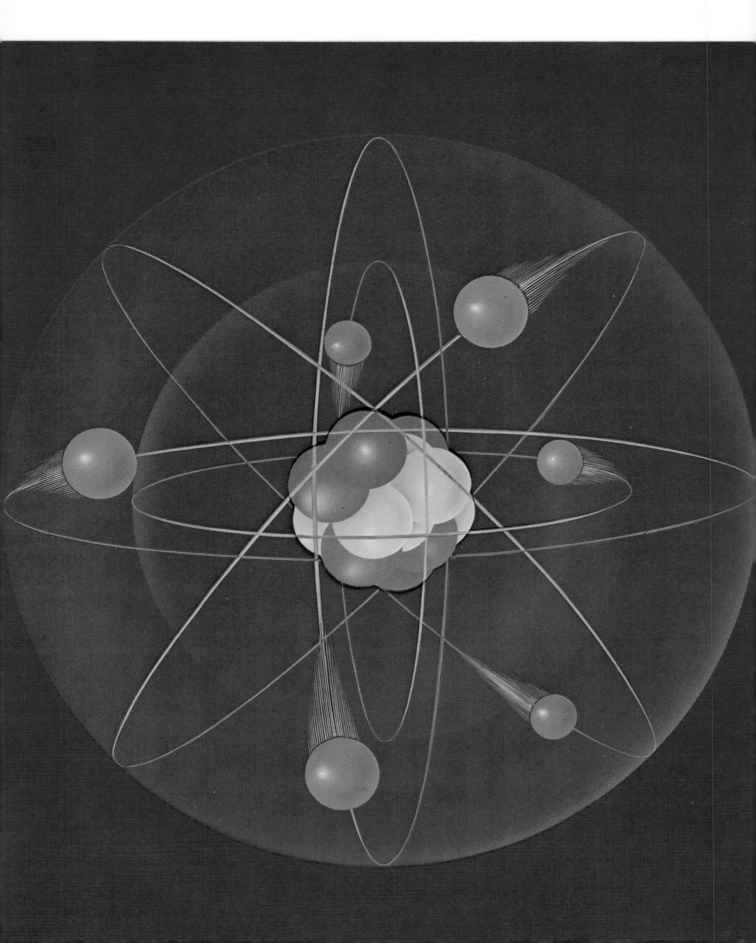

Chapter 1 Through Microscope and Telescope

How small is the smallest particle of matter? How big is the Universe? The Victorians thought they knew the answers. Then, in the last years of the 19th century, a series of exciting discoveries were made that revolutionized physics. The atom was cracked open and a new world was found within . . . one that also shed new light on events deep in space. But new mysteries arose, many of which remain unsolved. Are there still smaller particles locked up inside the subatomic particles whose existence we can now observe? Are parts of the Universe beyond the reach of our most powerful instruments? Will we ever know all the answers?

Areas of the night sky that to the naked eye seem to be empty and bleak are found to contain millions of stars when looked at through the world's most powerful telescope. Astonishingly, the light from some of these stars has taken many billions of years to reach us. And yet, no one suggests that even these distant stars are the farthest outposts of the Universe. What, then, lies beyond? Are there other worlds we cannot see and is there, eventually, an end?

An equally incredible world is revealed through almost any powerful microscope. In it, the structures of living and inanimate things can be viewed in astonishing detail. And by using electronic devices it has even been possible to see atoms—regarded for many years as the building bricks of the Universe. Still other techniques have revealed particles within the atoms themselves, and new particles continue to be discovered. How many more are there?

In recent years man's knowledge of the smallest and largest constituents of the Universe has grown at an astonishing rate. Some scientists hint that this new understanding of the two extremes—the microcosm and macrocosm—has brought us within an equation or two of the key to the Universe: a theory that would embrace all activity and influence both here on Earth and in the farthest reaches of space. But not everyone subscribes to that view. As old mysteries are solved new ones appear, encouraging some to believe that man can never know all the

Opposite: a representation of the structure of the carbon atom. It looks like an artist's impression of a planetary system in outer space, and it was precisely this image of the atom—the planetary theory—that formed the foundation on which modern physics has been built. The center, or nucleus, is made up of protons (particles containing a positive charge of electricity), and neutrons (uncharged particles). Around the nucleus spin electrons (negatively charged particles), which circulate in fixed orbits, much the same way as the Earth revolves around the Sun. The total positive charge in the nucleus exactly equals the total negative charge of the electrons, so the whole atom is neutral.

The Theory of the Atom

Below: Democritus of Abdera, the Greek philosopher who taught that matter was made up of very tiny particles, later called atoms. All chemical and natural changes, he believed, were associated with the combination and separation of these atoms —a theory that has earned him the title "the father of modern science."

Below right: according to Democritus atoms were of different sizes, shapes, and weights; they swirled around in empty space and combined to form all matter in the Universe. When living things such as trees and human beings die, or when inanimate objects such as rocks or stars disintegrate, their atoms were not destroyed but recombined to make new things.

answers. It would certainly be a sad day for mankind if the last mystery were ever to be unraveled and the last question answered.

The ancient Greeks were very good at asking profound questions about life and death, the workings of the Universe, and the nature of matter. Considering they had none of our sophisticated scientific tools, they were surprisingly good at coming up with the right answers, too. Over 2000 years ago, Greek intellectuals were debating whether or not matter could be divided and subdivided into finer and finer particles, without end. Two, Leucippus of Miletus and his pupil Democritus of Abdera, maintained that this was not the case. Ultimately, they said, tiny particles would be reached that would be indivisible. These particles Democritus named after their nature: *atom*, the Greek word for nondivisible.

Democritus then went one step further, arguing that different atoms existed, which were used to form different substances. By rearranging the atomic combination of a substance it could be changed into something else. This whole concept, put forward in 450 B.C., was not well received at the time. Plato, the famous Greek philosopher, and his equally well-known pupil Aristotle were among those who dismissed it. The idea of the atom was never totally forgotten, however. It emerged from time to time through the ages until at the beginning of the 19th century John Dalton, an English chemist, used an atomic theory to explain the difference between certain elements—and he gave credit where it was due by using Democritus's name, atoms, to describe the particles. Other scientists followed suit so that by the 1860s the theory had won general approval and a list of atomic weights for the various elements had been calculated. But not everyone accepted the idea that atoms were real. The skeptics were won over when it was demonstrated that molecules—groups of atoms —in water caused suspended pollen grains to jiggle perceptibly. The mystery of matter had almost been solved, it seemed, for though the atoms and molecules could not be seen, their influence was unmistakable.

In 1955 a field ion microscope was invented in America capable of magnifying an object up to five millionfold. The image of a needle tip at that enlargement shows the atoms as little bright dots. Later, the technique was improved to obtain pictures of a single atom. By 1970, an American physicist had detected individual atoms of uranium and thorium by using a scanning electron-microscope.

Left: John Dalton, the British chemist and philosopher. Democritus's early theories on the atom were verified by Dalton's experiments, which showed that each element had its own kind of atom and each compound its own kind of compound atom. He laid the foundation of modern chemical thought with his development of the atomic theory in the early 19th century.

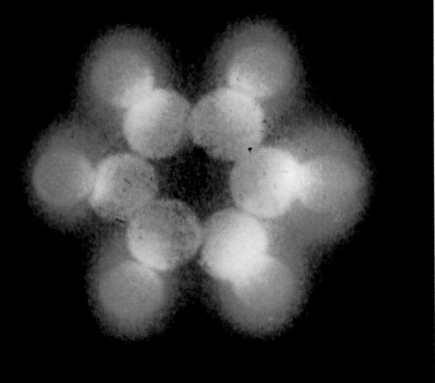

Left: a photomicrograph of the benzene ring, composed of six atoms of carbon each with an atom of hydrogen attached. Benzene was discovered by Michael Faraday, the English physicist, in 1825. Its structure—formulated 40 years later by August Kekulé, the German chemist— reinforces the Victorian view of compound atomic structures looking like minuscule billiard balls stuck together. The benzene ring is one of the basic molecular patterns of organic chemistry, and the discovery of its arrangement made possible later experiments leading to hundreds of compounds containing carbon.

Above: Ernest Rutherford, who worked out the nuclear theory of the atom in 1911. At that time there was much speculation on the way the atom was organized and arranged. Rutherford demonstrated that atoms were not solid structures, but consisted of a dense central part, the nucleus, surrounded by tiny, almost weightless particles of negative electricity called electrons. Remarkably, the atom appeared to consist mostly of space.

Right: Rutherford's experiments shown in diagrammatic form. He directed a stream of heavy, fast particles of radium called alpha particles at a thin sheet of gold foil (near right). Most of the particles passed through the foil but occasionally a particle would be deflected off course—as though it had hit something solid inside the atom of gold.

Far right: Rutherford concluded that the atom had a positively charged core that deflected the alpha particles. Those that met the core head on were turned back along their original path. This was the beginning of the nuclear theory of atomic structure: the idea that atoms consist of two or more portions separated by space.

Did this mean that Democritus was right? His view of matter, which visualized atomic structures looking like millions of minuscule billiard balls stuck together, was adopted and held by scientists until well into the 19th century. But by then science was moving rapidly ahead and new experiments were soon to shatter the neat concept of a world made up of billiard-ball-like orbiting bodies.

The breakthrough began with an experiment first suggested by Michael Faraday, the English physicist, which revealed that if an electrical discharge was passed from one electrode to another in a sealed vacuum tube a green glow appeared on the tube wall. (Faraday himself had not been able to achieve a sufficiently good vacuum to produce the effect.) Other experiments by Faraday enabled James Clerk Maxwell to evolve his electromagnetic theory, which revealed the true nature of light as an electromagnetic disturbance in the form of waves. Then in 1854, Heinrich Geissler, a German glassblower, perfected a vacuum tube

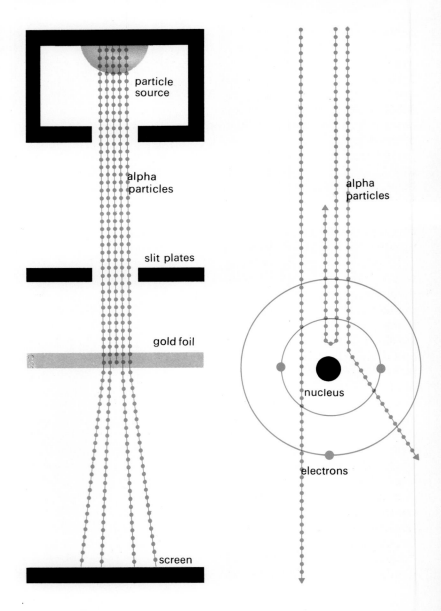

particle source

alpha particles

slit plates

gold foil

alpha particles

nucleus

electrons

screen

Inside the Atom

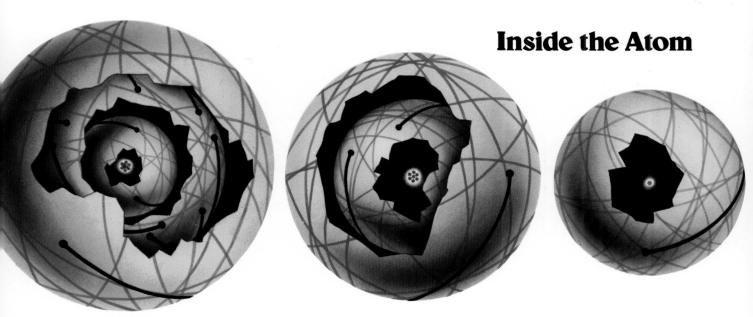

containing the electrodes that produced the green glow that so fascinated the Victorian scientists. The glow, it was suggested, was a form of radiation and was produced by "cathode rays," because Faraday called the negative electrode that appeared to produce the glow a cathode. But what were the mysterious cathode rays? Were they a form of the electromagnetic radiation described by Maxwell, or did the electric current create a stream of particles?

Experiments in 1897 by Joseph Thomson, a British physicist, established that the rays were streams of electrically charged particles. Although Thomson called them "corpuscles," by the turn of the century they had generally become known as *electrons*, and science had to face the exciting fact that they might be smaller than atoms. Also, under certain conditions, these small, or subatomic particles could be removed from atoms.

Proof that atoms were made of other particles came in 1911 when Ernest Rutherford, also a British physicist, was able to demonstrate that the atom was made up of a tiny nucleus, which contained most of its mass, surrounded by much lighter electrons. Then the atomic nucleus itself was found to consist of two varieties of particles, protons and neutrons. As far as scientists could tell, the extraordinary fact was that the atom contained more space than matter. The electrons that surrounded the nucleus of the atom helped prevent atoms from colliding, it was discovered, but they also enabled different elements to combine. A picture emerged in which electrons were seen to be orbiting in layers, or shells, around the nucleus. The negatively charged electrons nearer the center were very strongly attracted to the positively charged proton of the nucleus. Those in the outer shell felt less influence and were capable of merging with the outer shells of compatible atoms to form new elements. Atoms that had a full complement of electrons in the outer shell were stable. This was because the electrical charge of the layers of electrons just balanced the charge of the nucleus. Those with one or more "missing" electrons took the first opportunity to combine with similar atoms in a way that would bring about the right electrical

In the theory of atomic structure, the nucleus of an atom determines the identity of the element. Differences between nuclei distinguish one element from another.
Above right: the hydrogen atom, which has a nucleus of one proton and one orbiting electron.
Above center: lithium has a nucleus of three protons and four neutrons; two electrons orbit in an inner shell, one in an outer shell.
Above left: sodium, with 11 protons in the nucleus, has two electrons orbiting in the inner shell, eight in the second shell and one in the outer shell.
Below: uranium, the heaviest naturally occurring element, has a nucleus of 235 or 238 protons, and 92 electrons orbiting in seven shells.

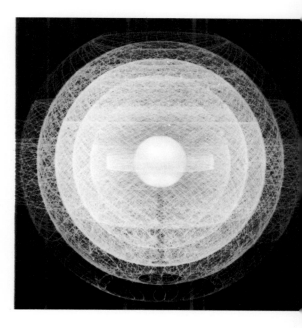

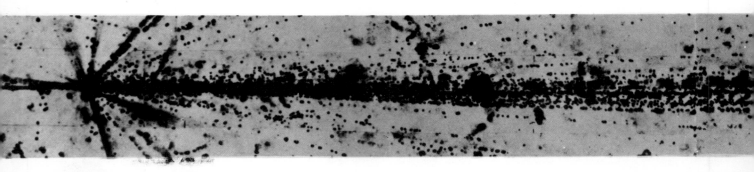

Above: a cosmic ray particle smashes into a nucleus of silver or bromide. Both the ray and the nucleus have been shattered into many different rapidly moving particles, and the track left by the collision has been captured on the emulsion of a photographic plate carried by a balloon some 9000 feet into the atmosphere. Knowledge of the components of atoms was advanced considerably as a result of research into cosmic-ray particles.

Below: Paul Dirac, the British physicist who predicted the existence of a new kind of particle—one with the same mass but opposite electrical charge to an electron. About two years later another scientist, the American Carl Anderson, was observing the behavior of cosmic ray particles when he discovered the antielectron predicted by Dirac. Antimatter had become a fact.

balance in all of them. Throughout the Universe atoms give up, accept, or share electrons in order to complete their outer shells.

There were still many unexplained mysteries about the nature of matter and to help make the picture clearer scientists began to explore new methods of splitting the atom into its various components. One method was to fire particles at dense substances in the expectation that some, instead of passing through the "empty" atoms like a meteor through space, would make a direct hit on the nucleus of an atom, breaking it down into more basic particles. A meteor racing toward Earth leaves a track in the atmosphere as it burns up. A subatomic particle passing through a chamber of cold, super-saturated air leaves a vapor trail. A study of these cloud-chamber trails taught experimenters much about the behavior of particles. By using magnetic fields to influence the particles they were able to determine if they were positively or negatively charged and the curve of the tracks indicated their mass and energy. The more violent the collision, the greater the chance of breaking particles down to their simplest form. But were they seeing *all* the subatomic particles? Some, it later turned out, were too short-lived to leave traces in the cloud chambers but they showed up in the far more sensitive bubble chambers invented in the 1950s, and researchers soon had thousands of photographs of new particles to study.

Although they did not solve the mystery of matter, the chamber experiments revolutionized man's understanding of the microcosm. Some of the discoveries were startling: one involved the study of the mysterious "cosmic-ray" radiation coming from outer space—first detected by an Austrian physicist in 1912 during high-altitude balloon flights. All scientists knew was that when the high-speed cosmic rays reached our atmosphere they collided with the atoms and molecules of the air, smashing the atoms into two or more "secondary" rays. During experiments in 1932 to find out the true nature of the cosmic rays—whether they were electromagnetic radiation or particles—Carl Anderson, an American physicist, passed cosmic rays through a cloud chamber. To slow down the rays so that he could detect a slight curvature more easily, Anderson placed a quarter-inch thick lead barrier between the rays and the chamber. It was then that he made his astonishing discovery. As they smashed into the lead they dislodged particles of the metal. These particles also left tracks and he noticed one that looked just like an electron. But there was a difference. It curved the wrong way. Its electrical charge was the opposite of an electron: it was in fact an anti-electron, which Anderson named a *positron*. The physicist had

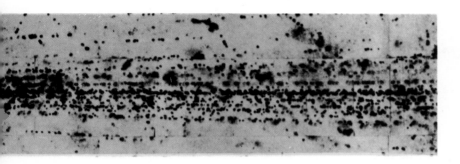

The Search for Antimatter

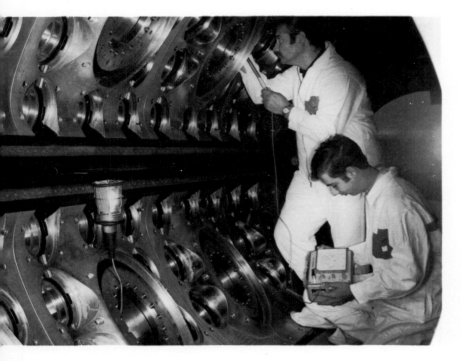

Left: the interior of a bubble chamber in the 12-nation nuclear research station near Geneva, Switzerland. The large holes are for cameras, the small holes for pressure control of the special heavy liquid with which the chamber is filled. Atomic particles are speeded up in an accelerator (to give them something of the character of the fast-moving cosmic rays), and are then released into the chamber. The particles speed through the liquid smashing into atoms and releasing electrons. Lines of bubbles trace the paths of the particles and their reactions with the atoms.

Below: one of the photographs of a bubble-chamber experiment. Fast moving particles entering from the bottom of the photograph have created electron and antielectron pairs (in blue). Other tracks are caused by stray cosmic-ray particles. The black-centered white blobs are lights used to illuminate the tracks.

discovered antimatter. In doing so he was confirming a prediction made two years earlier by Paul Dirac, a British theoretical physicist. After a mathematical analysis of subatomic particles, Dirac said antimatter should exist. According to his theory, when a particle and its antimatter counterpart meet they cancel each other out, leaving no matter at all, just a burst of energy. The discovery of antimatter also confirmed Albert Einstein's theory that matter could be converted into energy and vice versa. Because we are living in a world of matter, antiparticles cannot survive. They vanish in fractions of a second, annihilated by their "reflections" in the world of matter.

Scientists then turned their attention to the study of the atom nucleus, but they had first to develop the right kind of apparatus for such research. To disintegrate the nucleus of an atom meant bombarding it with high-energy particles at great speeds. A means was needed to accelerate the particles, and man in his quest for knowledge did just that. Various high-voltage particle accelerators have been developed and governments are now pouring billions of dollars, pounds, and roubles into building bigger, better, and faster accelerators. Particles can be accelerated by an alternating electric field that gives them successive

Subatomic Particles

Below: SPEAR, the electron annihilator ring at Stanford University in the United States. From the 2 mile-long electron accelerator, which leads into the square building at left, one stream of electrons creates antielectrons that are fed into the ring building on the right of square building. Another stream of electrons is fed into the opposite side of the ring and the two streams, which remain separate for most of the orbit, are made to collide and annihilate each other. The energy they release creates new particles—which either confirm theories already made, or create new puzzles.

"pushes." One of the best methods so far developed involves huge doughnut-shaped tubes that use magnets to deflect particles from their normally straight paths and keep them rushing around in circles, picking up speed and energy as they go. Einstein's theory of relativity predicted an increase in mass as an object accelerates and this has been confirmed in the particle accelerators. In America, accelerated protons have been endowed with so much energy that they have become 500 times heavier.

What has this intense research so far revealed? From it, and the work of brilliant theorists, an exciting new concept of the atom has emerged. The protons and neutrons of the nucleus are no longer seen as basic particles, for they too are made up of further particles. These, scientists believe, may be the real building bricks of the entire Universe. Further than that, by taking the atom apart, scientists have also learned much about the cosmic forces that shape our Universe. Forces such as gravity, electromagnetism, the strong nuclear force, and the weak force.

Gravity exerts its influence on all particles. It helps keep the Earth orbiting the Sun, the Moon and the various man-made satellites orbiting the Earth, and stops us from falling from the

face of our planet. The strength of the attraction depends on the number of particles in the masses and also on the distances between the masses—the larger the mass the greater the attraction but the greater the distance between two masses the smaller the attraction between them. The electromagnetic force is far stronger, at an atomic level, than gravity. It is this force that creates an attraction between protons and electrons, and between atoms and molecules. It is caused by the law of attraction between opposite electric charges. This law does not operate at the heart of the atom, however. There, only the protons have a charge, the neutrons are neutral. Normally this would lead to repulsion, but according to Hideki Yukawa, a Japanese physicist, a binding, strong nuclear force now comes into operation, exerting a far greater influence over these particles than that of electromagnetism. It forms such a strong bond that two million electron-volts are needed simply to disrupt the proton and neutron in the nucleus of a deuterium atom, one of the most weakly bound nuclei. Its strength is calculated at 130 times that of electromagnetism. But the field of influence of the strong nuclear force is incredibly small, falling to almost zero at a distance of just a

Above: Hideki Yukawa, the Japanese nuclear theorist who in 1934 first described the strong nuclear force that bound the protons and neutrons of the atomic nucleus together. They were, he said, short-lived force-carrying particles that shuttled between the particles of the nucleus and bound them together. Yukawa's strong nuclear force particles were discovered in cosmic rays in 1947, and since then many other kinds of subatomic particles have been discovered. The majority of the new particles appeared to be heavy, more energetic but very short-lived relatives of the proton of the nucleus. In a burst of energy they shed some of their mass and changed into protons.

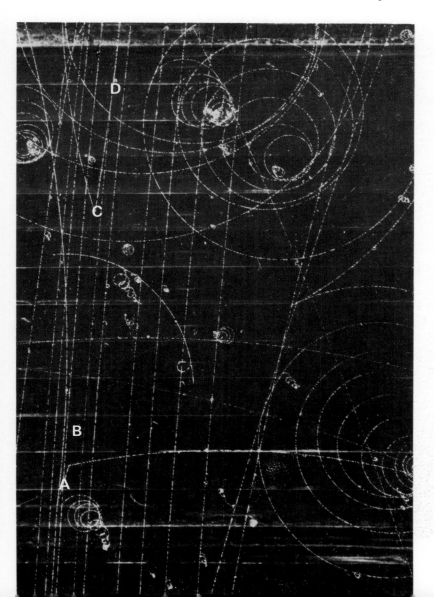

Left: a bubble-chamber photograph of one of the heavy particles that sheds mass and energy in three stages to become a proton. This particular particle is known as a "strange" particle, because it possesses the new quality of matter known as "strangeness"—a quality that affects the lifetime of particles. Depending on which stage they are away from the usual proton stage, a particle is known as strange, doubly strange, or omega-minus, which is triply strange. In the photograph the stages are marked A (for the omega-minus stage), B (doubly strange), C (strange), and D (non-strange proton).

Earth-centered Universe?

Below: the Moon, photographed with a fixed camera at 15-minute intervals. The Moon's apparent motion, revealed in the picture, is due to the Earth's rotation on its axis and not to the real motion of the Moon To the ancient astronomers, all the heavenly bodies were seen as satellites of the Earth. In fact, the Moon is the only true Earth satellite.

diameter outside a nuclear particle. Next is the weak force, whose subtle influence at the very heart of the atomic nucleus is capable of altering the quality of matter by ridding nuclear material of about half its electrical charges, thus enabling complex elements to form.

It must seem at times, even to the physicists involved in the fascinating subatomic exploration, that there is no end to the particles and forces that make up this minuscule world of matter. The same can be said of the other end of the scale. Is there, perhaps, a parallel between the concept of an atom as a nucleus with orbiting electrons and our own Solar System with the Sun and its orbiting planets? The microscopic solar system is not the same as the macroscopic one, and the forces at work on each are not the same either. But physicists hope that one day they can draw together all the cosmic forces in a theory that unifies them and depicts them as variants of one great force. Already a connecting link has been found between the weak force and electromagnetism and it may not be long before man can demonstrate that they are not, after all, independent forces. That achievement is just as likely to come from studying the stars as it is from plotting the tracks of invisible particles. It is even more likely that such a unifying principle—if real—will evolve through the joint efforts of particle physicists and astronomers.

Astronomy is largely an observational science, but since things are not always what they appear to be it has taken the application of advanced scientific techniques to give us a true picture of the Universe. Even so, there is still much about the heavens that mystifies us. The early astronomers could see that the Sun and stars revolved around the Earth. It was as obvious as the fact that the Earth was flat. Years of observations, however, planted seeds of doubt in the minds of the more adventurous thinkers. Philolaus of Tarentum, a Greek philosopher, was the first person in recorded history to suggest that the Earth was not flat but was a sphere. He made this suggestion in 450 B.C., the same year in which a fellow Greek, Democritus, put forward the theory of atoms. About 200 years later, another Greek philosopher, Eratosthenes of Cyrene, observed that there was a seven degree difference in shadows cast by the sun at the same time on the same day in places 500 miles apart. He calculated that if the Earth were a sphere and a seven degree angle represented 500 miles on its surface, then the planet's 360 degree circumference must be about 25,000 miles and its diameter about 8000 miles. This seemed so large that other astronomers made their own observations and reduced the Earth's size, making its proportions more readily acceptable. But Eratosthenes has been proved right. We now know our planet's circumference is 24,902.4 miles and its radius (allowing for slight variations since the Earth is not an exact sphere) is 7917.78 miles. Clearly, if we approach the mysteries of the Universe in the right way, even such things as shadows on the ground can reveal a tremendous amount about our world and others.

Despite the remarkable accuracy of Eratosthenes' Earth dimensions the ancient Greeks made little progress in drawing up a realistic model of the Universe. Hipparchus of Nicaea, the most famous of ancient Greek astronomers, did calculate that

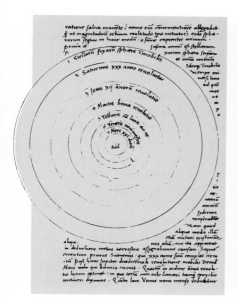

Left: Nicolas Copernicus, the 16th-century Polish astronomer. He was the first astronomer seriously to challenge the ideas summarized by the 2nd-century Alexandrian astronomer Ptolemy. According to Ptolemy, the Earth was the center of the Universe—an idea that held sway in Europe for more than 1000 years.

Far left: the Copernican view of the Solar System, with the Sun at the center and the planets moving around it. Copernicus was right about the Sun being at the center of the system, but erred in believing that the orbits of the planets around the Sun were circular.

Below: a 17th-century representation of the planetary orbits. Ideas of an Earth-centered Universe with all the planets moving in circular orbits were finally discredited after other astronomers improved and built on the work of Copernicus.

The Planets in Their Orbits

Below: Johannes Kepler, the German astronomer, represents the break between ancient and modern astronomy. He proved that not only did the Earth move around the Sun, but also that the planets moved in elliptical orbits according to precise mathematical rules.

the Moon was separated from us by a distance equal to 30 times the Earth's diameter. Using Eratosthene's figure of 8000 miles as the planet's diameter, the distance that separates us would be 240,000 miles. The average distance, we now know, center to center, is 238,854.7 miles. But other Greek astronomers calculated that the Sun was no more than 5,000,000 miles away (as against 93,000,000 miles in reality) and they still saw the Earth as the center of the Universe, orbited by the Sun and planets.

The first man to rethink the Sun-Earth-planet model was Nicolas Copernicus, a Polish astronomer. In a book published on the day he died, in 1543, he set out his reasons for believing that the Sun was the center of the Universe and that we and the other planets orbited it, in perfect circles. By 1609, Johannes Kepler, a German astronomer, decided that the planetary orbits were not perfect circles but were flattened into ellipses. Man, then, was slowly stumbling toward a better understanding of the nearest objects to him in the sky. But whereas our Solar companions' paths could now be understood, the stars just twinkled and remained in their positions year in and year out, giving no clue about their real size or distance from us. They looked like small lights stuck on a solid black sky, and that is precisely how they were regarded until less than four centuries ago. True, some

medieval scholars had suggested that they might be spread out through a vast space, but there was no evidence to support this hypothesis. Not, that is, until Edmund Halley, the English astronomer (famous for the comet which is named for him), noticed a surprising discrepancy between maps of the stars made by the Greeks—notably Hipparchus who recorded the positions of 800 stars about 134 B.C.—and his own observations of the night sky in 1718. Three stars, Sirius, Procyon, and Arcturus had moved perceptibly from their former positions. Why? It was obvious to Halley that no error had occurred. The stars really had moved during the 1800 years that had elapsed between the two observations. Man was aware at last that the stars, like the planets, were independent bodies in motion. But were they all at the same distance from the Earth? It was only 140 years ago that we found a way of solving what had always been one of the great problems of astronomy.

Right: the apparent movements of the planets as measured without modern instruments. Their complexity baffled ancient astronomers because they could not be reconciled with the idea of uniform, circular orbits around the Earth. Below: a photograph of the planets when plotted against the background of fixed stars. Kepler's major breakthrough was in plotting planetary movements.

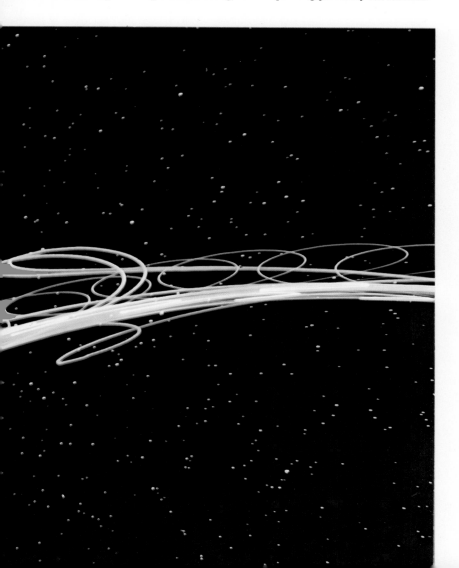

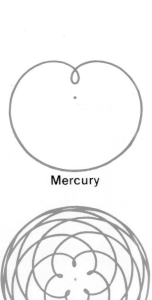

Mercury

Venus

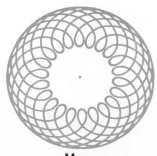

Mars

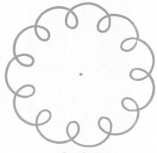

Jupiter

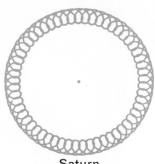

Saturn

Mapping the Stars

At an observatory in South Africa in 1831, Thomas Henderson, a Scottish astronomer, carefully plotted for a year the position of Alpha Centauri, the third brightest star. In the Baltic provinces of Russia, Friedrich von Struve, a German astronomer in the service of Czar Nicholas I, was plotting the position of Vega, the fourth brightest star. Almost simultaneously Friedrich Bessel, another German astronomer, was in East Prussia measuring the distance between 61 Cygni and two dim neighboring stars. All three were attempting to measure the *annual parallax* of the various stars: That is, if a star is nearer to Earth than others it will appear to change its position in relation to more distant stellar objects as the Earth travels around the Sun. If a near star is plotted for a year it will produce a small closed ellipse in the sky whereas stars that are more distant will not move perceptibly. The three astronomers all succeeded. Bessel was the first to announce his results in 1838. He discovered that 61 Cygni was 64,000,000,000,000 miles away. Because distances are so vast in space we now talk in terms of how long it takes light—traveling at 186,282 miles a second—to reach us from a star. It travels 5,880,000,000,000 miles in a year, so we describe that distance as a light-year. The star 61 Cygni, then, was about 11 light-years away. Two months later Henderson revealed that Alpha Centauri was 4.3 light-years away. Finally, in 1840, Struve published his findings. Vega was 27 light-years away. The Universe was on a far grander scale than anyone had imagined.

We now know that Alpha Centauri—not in fact a single star, but a cluster of three—is our nearest stellar neighbor. To put this in perspective against our Solar System, the farthest planet from the Sun is Pluto which is 3,666 million miles away. If you were to draw a large circle with a radius of nine yards to represent the Sun at the center and Alpha Centauri's position on the circumference, then Pluto's orbit around the sun would have a maximum diameter of one tenth of an inch.

Man was at last beginning to grasp the astonishing size of the Universe and to ask some very penetrating questions. Since then, huge telescopes have peered deeper into space. New techniques have been discovered to analyze the components of stars. Far more is known about the structure of the Universe and the way in which stars and planets are formed. But the questions persist. When and how was the Universe created? Does it have a beginning and an end? How old is the Earth? How long can life survive on this planet? Are we alone in the Universe, and if not then what are the chances that extraterrestrials have visited us in the past or are doing so now?

As well as answering these questions in the light of modern scientific knowledge, the following chapters also examine some very remarkable and perplexing objects in space that confirm the view that, despite the tremendous knowledge we have acquired in the last century, the Universe is still a very mysterious place.

Opposite: a painting of around 1700 by the Italian painter Donati Creti. It shows astronomers of the time using their instruments to observe Jupiter. Jupiter itself appears as their telescopes would then have revealed it, with red spot and satellites clearly visible.

Above right: Friedrich Wilhelm Bessel, the German astronomer. He was one of the three astronomers who first succeeded in plotting the distance of stars.
Right: the instrument first used to measure star distance was the heliometer— so called because it was originally designed for solar measurements. Using it, Bessel discovered that 61 Cygni was 11 light-years away.

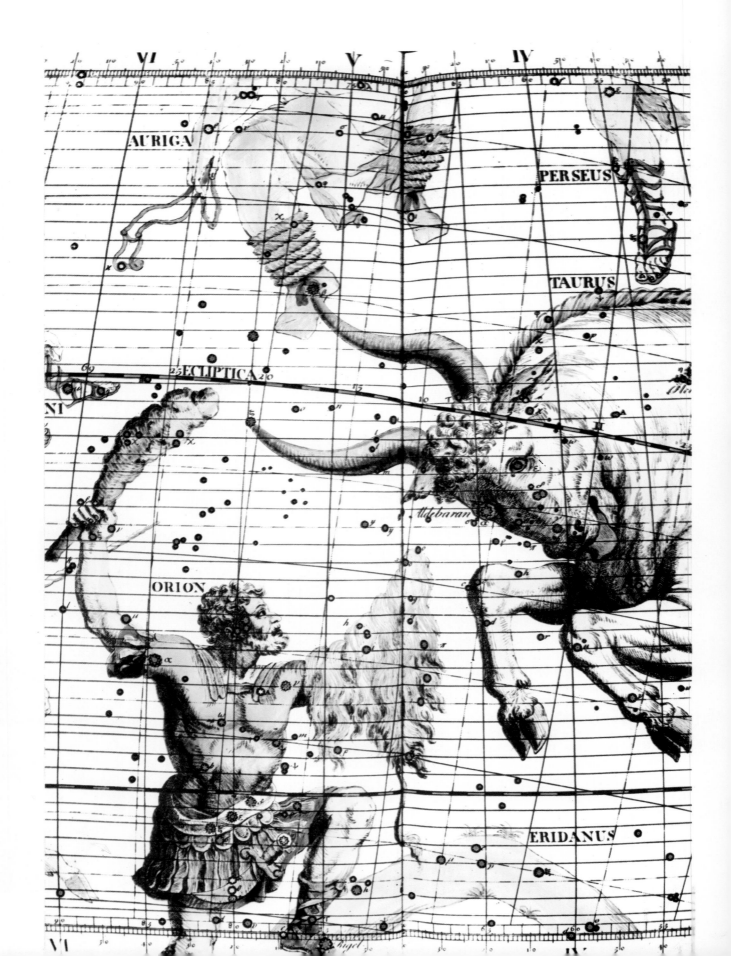

Chapter 2
The Architecture of Space

Ever since our early ancestors first gazed into the night sky we have been intrigued by the brilliance and mystery of our starry canopy. The deeper we probe into space the more mysteries we uncover. Some stars blink in a regular way, others suddenly flare up for short periods, and there are others that blow themselves apart in explosions of incredible magnitude. In our thirst for knowledge we have reached the Moon and probed as far as Mars and Venus. Is star travel now a possibility? Are we on the threshold of an exciting new era in our understanding of the cosmos?

Long before man had even a vague concept of the majesty and sheer size of the Universe, he was able to derive practical knowledge from its behavior. To the early Egyptian civilization, the annual flood of the Nile river was an event of great importance. The four-month inundation fertilized the barren soil enabling the farmers living along its banks to cultivate the land. But how did they predict the flood with accuracy? Their early calendar, based on the Moon's movements, was not accurate enough to be able to pinpoint the great river's fluctuations. They solved the problem when an observant stargazer noted that the flood always came when the star Sirius first became visible in the dawn sky. This discovery also enabled them to draw up a calendar based on a year of $365\frac{1}{4}$ days.

The Egyptians were not the only ancient civilization to notice the regular rhythmical movements of Sun, planets, and stars across the sky. The Babylonians were great astronomers, as were the Hindus and Chinese—in fact, the Chinese had established a 365-day calendar around 3000 B.C. Mysterious stone monuments erected before recorded history—such as Britain's famous Stonehenge—seem to bear silent testimony to a good working knowledge of the Sun and Moon's movements. They may well have been incredibly accurate observatories with which ancient man was able to predict important lunar events and even eclipses of the Sun.

Without understanding why, early man realized that the move-

Opposite: part of a celestial map from the *Atlas Coelestis* in the British Museum. When ancient astronomers looked into the night sky they sought to put into some kind of order the glittering canopy of stars above them. One group of stars reminded them of a hunter, another the outline of a mighty bull, so they named the patterns for the heroes of their myths and the gods of their religions.

Above: part of a Babylonian record of positions of Jupiter composed during the the 1st and 2nd centuries B.C. Ancient astronomers had neither the knowledge of the universal laws of gravitation, nor highly developed optical instruments. Their observations of the Sun, Moon, planets, and stars were therefore concerned with noting and trying to interpret the apparent motions of the heavenly bodies.

Right: The constellation Perseus as depicted by a 16th-century Persian astronomer. For many centuries people have grouped the stars into patterns that reminded them of their heroes or of mythical or real animals.

ments of our neighbors in space followed an orderly pattern, and one that helped him to live an orderly life if he allowed the Sun, Moon, planets, and stars to guide him. Not surprisingly, then, he endowed these bodies with special powers. They influenced not only the seasons and the tides but man himself, he believed, and so *astrology*—which sought to interpret the relationship between the heavenly bodies and human beings and predict man's destiny in terms of the movement of planets and stars— emerged, and has continued to influence man for over 3000 years.

The Babylonians were responsible for discovering and giving names to the 12 signs of the zodiac and attributing good and bad qualities to the planets, as well as interpreting their *aspects*— the angles formed between planets. The Egyptians eventually took to astrology and used signs of the zodiac to decorate the tombs of some of the pharaohs. Then the new "science" spread to Greece where it was adopted enthusiastically, and modified with new interpretations.

It was the Greeks who made personal horoscope casting popular and it is a fashion that persists. Yet despite its widespread appeal, astrology is dismissed by most astronomers today as no more than a superstition. In the early days of man's stargazing, however, it was impossible to draw a fine distinction between the two. The pioneer astronomers were astrologers, and their studies of star and planetary movements helped lay the foundation for modern cosmology. Around 200 B.C. Claudius Ptolemy, regarded as the greatest astronomer of ancient times, produced a number of manuals on astrology—the most important being the *Tetrabiblos*. His catalogs and star maps were not rivaled for their accuracy until the 17th century. Ptolemy's view of the Universe placed the Earth at the center with the stars and all other bodies in a fixed sphere, revolving around it.

Many of the great 16th and 17th century astronomers were also practicing astrologers—such as Johannes Kepler, the German who was the first to prove conclusively that the Earth went around the Sun. Sir Isaac Newton, the English scientist whose laws of gravity are said to have been inspired by an apple falling on his head, originally took up mathematics in order to practice astrology. And when the first English Astronomer Royal, John Flamsteed, had to arrange the date for the laying of the foundation stone of the Royal Observatory at Greenwich, he used a form of astrology—of which he was a staunch advocate—to choose August 10, 1675.

During the 18th century, however, astrology suffered a serious decline in Europe and almost died out. Since then it has enjoyed periods of revival, but with the comparatively recent explosion of knowledge about our Solar System and the stars that form

From Astrology to Astronomy

Below left: the observatory in Prague where Kepler did much of his work. Many of the advances in astronomy in the 16th and 17th centuries came through men who, like Kepler, were also practicing astrologers. Because astrologers foretold a person's future from a knowledge of the positions of the stars and planets, they had to be skilled astronomers, too.

Below: Isaac Newton, the British physicist and philosopher. It is said he studied mathematics in order to practice astrology. He went on to discover the laws of gravity— laws which indirectly removed much of the mystery surrounding astronomy and made humans wonder whether the planets and stars—which appeared to be subject to the same laws as the Earth—could any longer be viewed as rulers of their destiny.

such a brilliant and dramatic backdrop to events played out on our tiny planet, astronomy has left little room in the scientific mind for the mystical aspect of the Universe.

Meanwhile, there was still much in the heavens that mystified astronomers of the 18th century. Charles Messier, a French astronomer with a particular interest in comets (he discovered 13), became very frustrated when many of the cometlike objects he thought he had discovered proved time and again to be something quite different: for unlike comets they did not move. In order to avoid confusion Messier, with the aid of other astronomers, made a list of these objects, which eventually totaled 100. Each fuzzy patch of light was given a number, preceded by the letter M—for Messier. But Messier's instruments were not powerful enough to discern anything more than a patch of light, and

Left: Charles Messier, the French astronomer who compiled a list of hazy, nebulous patches and galaxies, which he published in 1784. By listing these stationary objects, he hoped to avoid confusing them with comets, which also look like hazy patches before they grow a tail.
Below: William Parsons, the Earl of Rosse. The 72-inch reflector telescope shown here was built by him and was for many years the world's largest telescope. With it he detected the spiral structure of many nebulae.

the mystery of these *nebulae* (from the Greek word for cloud) and star clusters had to remain unsolved for some time. William Parsons, a British aristocrat-astronomer, who devoted all his time to building a 72-inch telescope, noted in 1845 that some of the nebulae had spiral structures. This did not solve the problem of their composition, however, and speculation suggested that the nebulae were huge clouds of gas and dust.

Meanwhile, science was providing a new tool with which astronomers could extract information from the stars. In 1814 Joseph von Fraunhofer, a German optician, passed sunlight through a small slit and then through a prism to produce a spectrum. This was a refinement of Sir Isaac Newton's experiment that showed that light could be separated into a spectrum of colors. Fraunhofer was a superb glass technologist and his equipment pro-

Nebulae and Star Clusters

Above: the Whirlpool galaxy (still also known by Messier's catalog number, M51) as seen through Parsons' 72-inch telescope. Parsons made the drawing in 1850, after he had established its whirlpool shape.

Left: the Whirlpool galaxy as seen through a modern telescope. It was Parsons who first suggested that it was composed not of gas but of individual stars. The Whirlpool's spiraling arms can be seen here to contain many highly luminous stars. It is 10,000,000 light-years away and receding from earth at 300 miles every second.

Spectra of the Stars

Below: light being diffused through a spectroscope. By photographing the spectra of stars and other celestial bodies, valuable information on their composition and behavior is gained for astrophysicists. The instrument used here is known as a diffraction grating spectroscope, which is many times more efficient at separating color than the prism.

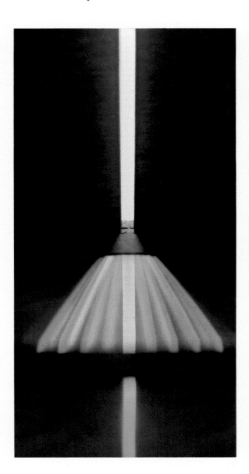

Right: Scientists gained their first insight into the compositions of the stars through the work of the early spectroscopists. In the 1860s, William Huggins, a British astronomer used this eight-inch telescope fitted with a spectroscope at the viewing end to obtain spectra. Spectra produced in the observatory showed that stars possess many elements in common with the planet Earth.

duced very sharp images. It was possible to note dark lines at some points in the spectrum, indicating that particular wavelengths of light were missing. Scientists immediately searched for an explanation. It soon became apparent that the lines were linked to elements in the Sun. Dark lines indicated absorption by certain elements whereas the bright lines represented elements emitting light. Two German chemists then devised a system of identifying the various elements and their spectroscope was soon used in the study of light from the Sun and stars. And so, in 1862, Swedish astronomer Anders Jonas Ångström was able to detect hydrogen in the sun and the stars. But in the same year a new puzzle appeared.

Alvan Clark, an American instrument maker, was testing a new telescope when he spotted a dim star near to Sirius. This proved to be the hitherto unseen companion around which Sirius

was known to be circling. The mystery was that although instruments told its observers that its mass was equal to that of our Sun it was only one-four-hundredth as bright. The only possible explanation at that time was that this must be a dying star, but Sirius B (as it became known) was to prove to be far more interesting. It was not a dying star, it was later discovered, in fact scientists knew it to have a high surface luminosity. The explanation is that Sirius B is only 25,000 miles in diameter, against Sirius's 1,500,000 miles. Whilst it is very hot, its dimness is due to its small size—too small to be visible to the naked eye.

Every so often through the ages new, bright stars appear. Sometimes they are so spectacular that they outshine the other lights in the night sky and are even visible in daylight. Tycho Brahe, the Danish astronomer, was so impressed with a new star he saw in 1572 that he wrote a book, *De Nova Stella*, (The New Star) and each new occurrence of the phenomenon has been known as a *nova* ever since. With the invention of the telescope it became obvious that the "new" stars were not new at all, just stars that for some mysterious reason suddenly brightened for a comparatively short period, then slowly dimmed to their former obscurity. From a study of novae it was realized that they were exploding

Above: starlight viewed through a spectroscope. Spectroscopy can give astronomers much valuable information about stars that are in most cases billions of miles away. They can tell, for example, a star's temperature, what elements it contains, what kind of magnetic field it possesses, and its speed of rotation. The spectra reproduced here are of stars from the Hyades open cluster, in the constellation Taurus. It is one of the most conspicuous star clusters in the sky and is about 135 light-years away.

What the Star Spectra Tell Us

spectral type and typical spectrum

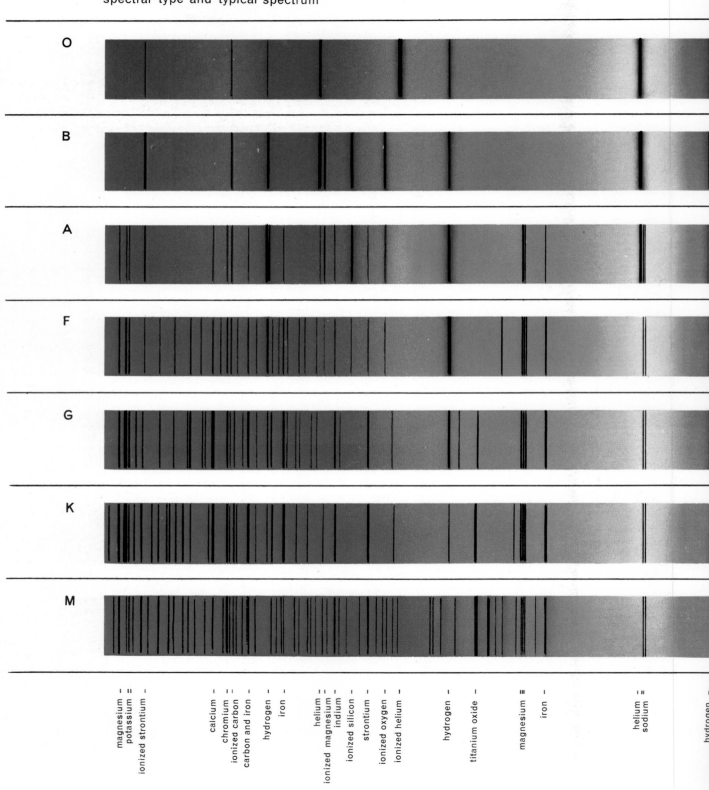

O
B
A
F
G
K
M

magnesium =
potassium =
ionized strontium —
calcium —
chromium =
ionized carbon =
carbon and iron —
hydrogen —
iron —
helium =
ionized magnesium =
indium —
ionized silicon —
strontium —
ionized oxygen —
ionized helium —
hydrogen —
titanium oxide —
magnesium ≡
iron —
helium =
sodium =
hydrogen —

atoms producing main lines of spectrum	color of main radiation	temperatures at surfaces of stars
ionized helium, neutral hydrogen and helium		35,000°–40,000°C
neutral helium; ionized silicon, magnesium, oxygen and nitrogen; neutral hydrogen		11,000°–35,000°C
metals (especially calcium) giving weak lines; hydrogen giving very strong lines		7500°–11,000°C
metals (especially calcium) giving strong lines, hydrogen giving fairly weak lines		6000°–7500°C
potassium giving a strong line; neutral metals (fairly strong lines); hydrogen (weak lines)		5100°–6000°C
neutral metals giving strong lines; hydrogen (very weak lines)		3600°–5100°C
molecules of titanium oxide (strong bands)		2000°–3600°C

Astronomers use letters (extreme left) to classify stars, with the brightest and hottest stars falling into the O group and the least bright and hot in the M group (our Sun, incidentally, is a G group star). Stars in these seven groups are known as main sequence stars. Next to the spectral type is the typical spectrum. When white light from a star is passed through a spectroscope the light is split up into its component colors. The colors are not continuous, however, but are interrupted by a series of dark lines. The lines always fall in the same place for each element so that scientists can tell just what elements a star contains. The intensity of the lines shows the predominance of the various elements. If, for instance, a star contains more hydrogen than any other element, more radiation will come from the hydrogen atoms. The hydrogen will therefore produce more intense lines than the other elements. As well as a star's composition, astronomers can calculate its temperature, speed and direction of motion, rate of rotation, and strength of its magnetic field, from the number, thickness, and position of the lines.

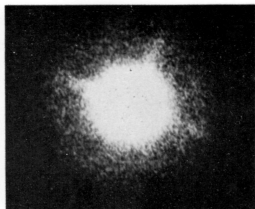
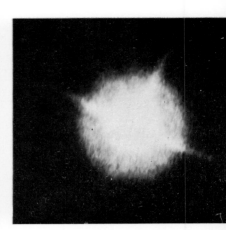

Above: three photographs of Nova Aquilae, which increased greatly in brightness in the year 1918. They show how its shell of gas looked in 1922, 1926, and 1931. Exploding novas occur when an abundance of hydrogen reaches the center of an aging star, causing a tremendous increase in the star's brilliance—up to 10,000 times more brilliant than an ordinary star. Radiation is blown off the surface of the star in a huge globe of very hot, bright gas.

Right: the Crab Nebula in the constellation Taurus is the remains of a supernova explosion observed by Chinese astronomers in 1054. Whereas a nova explosion involves only the outer regions of a star, a supernova involves the whole star being torn apart in a cataclysmic explosion—its brilliance reaching up to 10,000 million times the luminosity of an ordinary star.

stars, capable in some cases of increasing their brilliance a millionfold in a day. But the star was not exploding into tiny fragments or into a puff of gas. It ejected only one or two percent of its mass and then apparently returned to a comparatively normal existence.

In 1885 Ernst Harwig, a German astronomer, discovered a nova in the most spectacular of Messier's cloudlike objects— M-31, or the Andromeda nebula. Through a telescope the nova appeared to have a brightness one-tenth that of M-31—and nearly bright enough to be seen with the naked eye here on Earth. At that time astronomers did not know the Andromeda nebula's distance from Earth. Now that we have that knowledge it is clear that Harwig was watching a very rare event. The Andromeda

Supernovas, the Exploding Stars

Below: a brightness contour map created in the Kitt Peak National Observatory, Arizona, of a supernova discovered in 1975. The elongated red area on the right is a galaxy in Ursa Major, the round red area on the left is the supernova. The colors on the contour map are those used at Kitt Peak to indicate brightness: red for the brightest areas, blue for the dimmest. The supernova, it can be seen, is as bright as the whole parent galaxy.

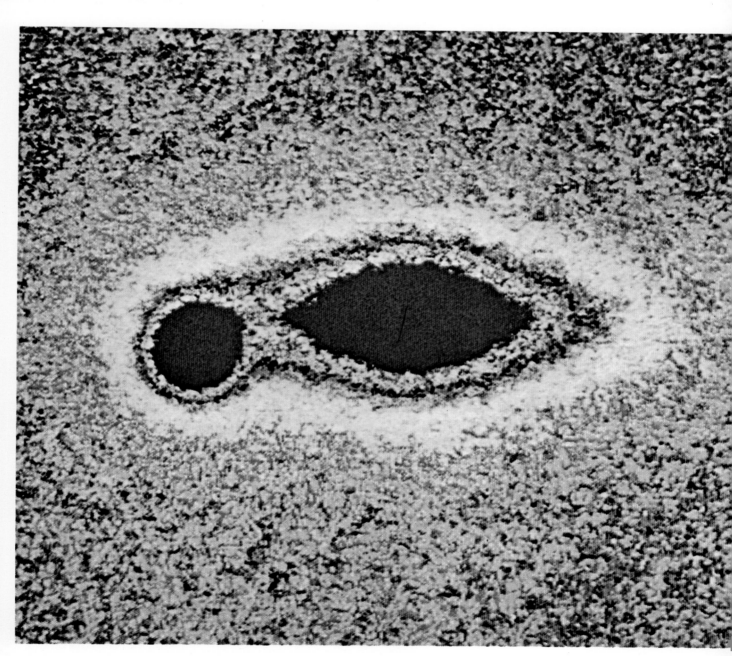

Evolution of a Supernova

Right: the evolution of a supernova depends on a star's mass. *A* represents a star only a few times more massive than the Sun.

1: hydrogen in its core is fused into helium. After several billion years the hydrogen is used up. 2: the star's core contracts, its exterior expands and the star becomes a red giant. 3: the outer layers of the red giant are eventually expelled and become a planetary nebula. 4: all that remains of the star is a white dwarf.

B: some supernovas are members of double-star systems. Stages 1 and 2 are the same as their solitary counterparts, except that as one star reaches the red giant stage it begins to lose matter to its companion. 3: the companion expands to a red giant, the other star becomes a white dwarf. 4: matter is suddenly transferred from the red giant and added to the white dwarf, which increases its mass beyond the critical limit of 1.44 solar masses. 5: the core of the white dwarf collapses, releasing energy as a supernova. 6: a binary system remains in the form of an ordinary giant star and an x-ray source (wavy blue arrows).

C: If a star is much more massive than the Sun, the evolution to supernova is different after stages 1 and 2. At stage 3, hydrogen in the red giant continues to be burned in a shell (red) around the core, and the core contracts and heats up until helium (yellow) fuses into carbon. At the next stage one of two catastrophes may overtake the star. When the helium is exhausted the core begins to burn the carbon. 4: if the ignition of the carbon (black) induces instabilities, the star explodes cataclysmically as a supernova (5) leaving behind only an expanding gaseous remnant (6). If, after stage 3, the carbon is safely ignited, extraordinarily high temperatures build up in the core. 7 and 8: the star throws off radiation particles (blue arrows) at an ever-increasing rate, sapping its energy so that its core plunges to total collapse. 9: a final burst of radiation might cause the red giant to throw off the outer envelope of the star in a gigantic explosion. It could leave behind a nebulous collection of gas, at the center of which would be either a pulsar, or (10) a black hole.

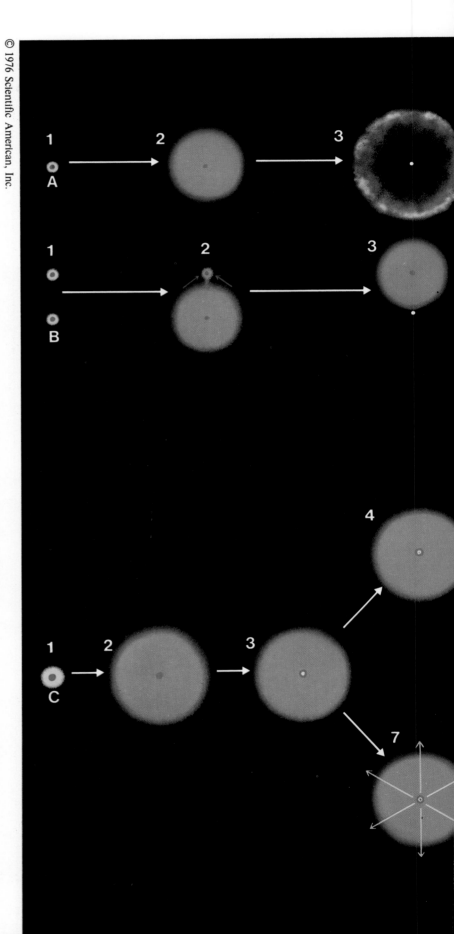

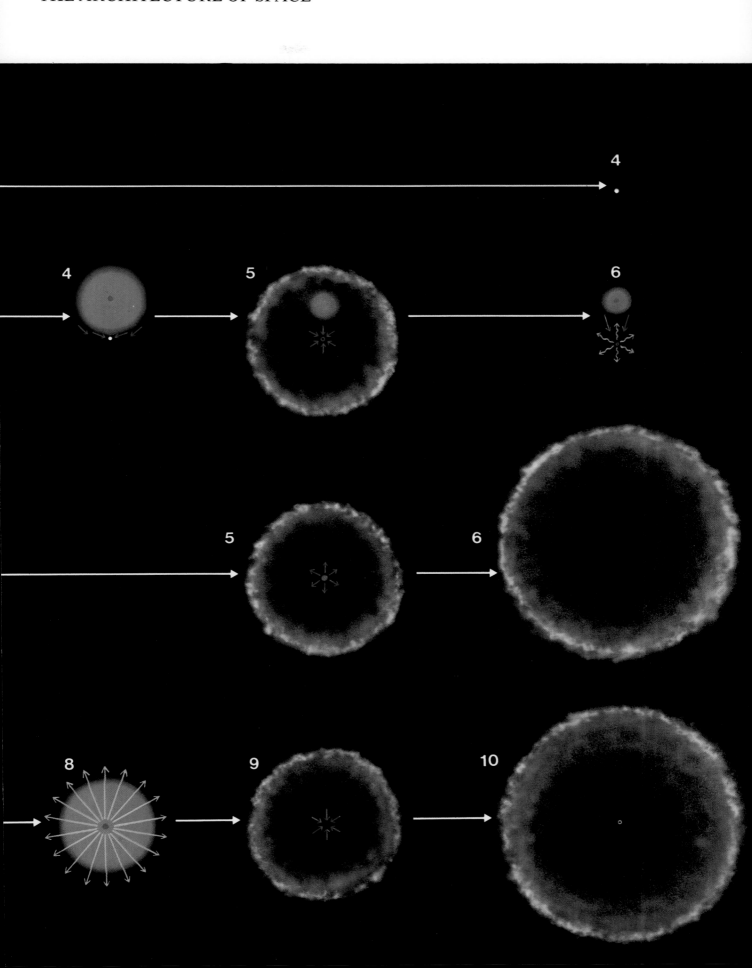

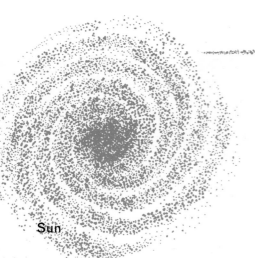

Sun

Sun

Above: the Earth's galaxy, the Milky Way, is believed to be shaped like a flattened disk that swells to a hub at the center. The Solar System is some 30,000 light-years from the center, while the whole disk is some 100,000 light-years across.

Left: the Milky Way seen from above. Its spiraling arms give it something of the appearance of the Whirlpool galaxy (page 29).

Below: a composite picture of the Milky Way galaxy. It is thought that the Milky Way was formed from a very large cloud of gas hundreds of thousands of light-years in diameter. Most of the gas reached the disklike form before stars were formed, and then star formation could have happened very quickly. The gas that remained uncondensed now forms the interstellar gas.

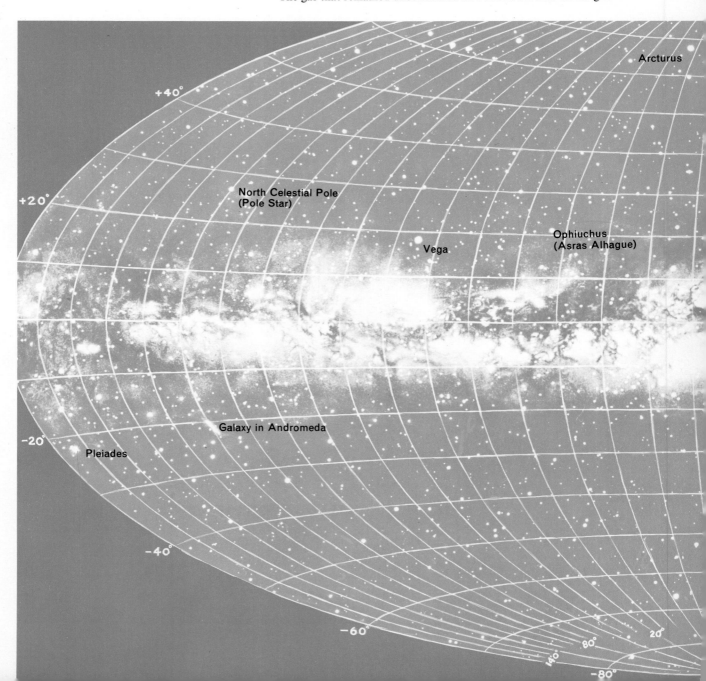

Arcturus

+40°

North Celestial Pole
(Pole Star)

+20°

Ophiuchus
(Asras Alhague)

Vega

-20°

Galaxy in Andromeda

Pleiades

-40°

-60°

-80°

flare-up was caused by a *supernova*. The difference between that and an ordinary nova, it seems, is that the supernova is blowing itself apart in an incredible explosion.

By 1900 the distances of some 70 stars had been determined out of the 6000 that are visible to the naked eye, and a far clearer picture of the Universe was emerging. One mystery that had been solved to most people's satisfaction was the reason for the ribbon of milky cloud that encircles our planet. When Galileo Galilei, the Italian astronomer, focused his telescope on this region—the Milky Way—he was astonished to find that it was made up of millions of stars. But why were they clustered together in such a band? The answer put forward in 1785 by William Herschel, the German-born astronomer, was that the *galaxy* (from the Greek word for milk) was shaped like a lens. If we look from our vantage

The Milky Way

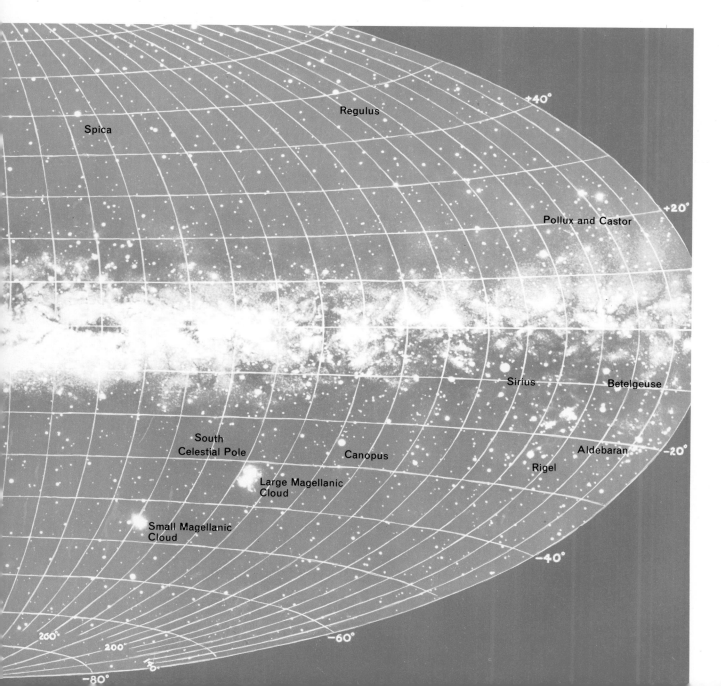

Redrawing the Star Map

Below: Henrietta Leavitt, the American astronomer who provided astrophysicists with a valuable measuring scale through her work on the varying luminosity of a particular type of pulsating star known as a Cepheid (named for the earliest studied prototype, Delta Cephei).

Below right: the Small Magellanic Cloud, from one of Henrietta Leavitt's original photographs used in her classic investigations into the short-period variable Cepheids. Because Cepheids are bright stars they can be observed at great distances. Also, their pulsating light is an easily distinguishable feature. This can therefore be used for measuring far greater distances than those obtained by observation of nonvariable stars. The Small Magellanic Cloud and its companion the Large Magellanic Cloud are the closest of the external galaxies—smaller companions of the Milky Way.

point on the thin edge of the lens in toward the center the stars will appear very numerous. We are, in fact, part of the ribbon of stars. Whereas if we look up through the top or down through the bottom (if we allow space to have a top and bottom) we see far fewer stars. A much more accurate picture of our galaxy was constructed in 1906 using not only a telescope but also photography to provide a permanent record of the Milky Way. This work was carried out by Jacobus Cornelius Kapteyn, a Dutch astronomer, and from his observations he calculated the galaxy's measurements as 23,000 light-years in diameter by 6000 light-years in depth. This was over four times larger than Herschel's original estimate, but we now know that even Kapteyn's measurements were conservative.

By 1925 Jan Hendrik Oort, another Dutch astronomer, was able to calculate the galaxy's rotation and determine that we are 27,000 light-years from its center, oribiting at a speed of about 140 miles a second on a journey that takes 230,000,000 years to complete one revolution. The galaxy was then established to be between 80,000 and 100,000 light-years across, 16,000 light-years thick at its center, and containing at least 100,000,000,000 stars. Vast clouds of gas and dust obliterate our view of the galactic center, which is why the Milky Way looks like a huge ribbon of light instead of a mountain or wall of stars and gas.

Twelve years before Oort's calculations, a United States astronomer made a discovery that provided a new yardstick with which to measure the vast distances to the stars. As far back as 1784 astronomers had noted that Delta Cephei, a bright star in the Cepheus constellation, had a cycle of brightness. From its

dimmest stage it would double in brightness very quickly then slowly dim, and it repeated this variation in its brightness over each 5.3-day period. Since then, other stars had been found that behaved in the same way and they are now known as Cepheid variables, or just Cepheids. The time it takes for a cycle to be completed varies from start to star. The shortest period noted is less than one day, the longest close on two months. Why did these periods vary and what was the significance of the Cepheids?

It was in 1913, while astronomers were trying to answer these questions, that Henrietta Leavitt, working at Harvard College Observatory, made her discovery. She was studying the smaller of the two huge star systems visible in the Southern Hemisphere known as the Magellanic Clouds, and named for Ferdinand Magellan because they were first reported during his famous voyage around the world. Miss Leavitt found 25 Cepheids in the Small Magellanic Cloud and her studies of these revealed that the longer the period of variation between brightness and dimness the greater was the star's luminosity—known as a star's magnitude. This remarkable fact had not been noted before because the various Cepheids studied had all been at different distances from the Earth and it was not possible to determine how far they were from us. But because the Cepheids Miss Leavitt had studied were all in the same star system, and therefore approximately the same distance from us, the link between variability and brightness was evident. Working on the assumption that all Cepheids in the Universe displayed this same behavior, Miss Leavitt was able to produce a "period-luminosity curve." In other words, if astronomers saw two Cepheids in different parts of the sky with the same variable periods, they could assume that they also had equal magnitude of light. If one looked brighter than the other, it was obvious that the brighter of the two was nearer to us than the other. If a Cepheid looked four times as bright as another with the same period, then the dimmer Cepheid must be twice as distant. With this information it was possible to calculate a scale map. All that was required then was for the exact distance of just one Cepheid to be established for the rest to be known. Eventually, after observations involving complicated procedures, another American astronomer, Harlow Shapley, succeeded in calculating the distances and in establishing Cepheids as a very useful measuring stick, giving a new perspective to space and a new understanding of many stellar phenomena.

At around the same time astronomers were showing renewed interest in nebulae, those luminous cloudlike objects that had been listed by Messier. Immanuel Kant, the German philosopher, had called them "island universes" in 1755, but there was no evidence that they were more than comparatively close bright clouds of gas and dust. Then, in 1924, when the American astronomer Edwin Powell Hubble put the Andromeda nebula under the searching gaze of the new 100-inch Mount Wilson telescope in California, he discovered individual stars in its rim. The nebula looked very much like our own galaxy and astronomers began to realize that elsewhere in the Universe were immense star complexes, some even bigger than our own galaxy which, for so long, we had regarded as the entire Universe. Far from being the center of the Universe, the Earth was now seen to be a small planet

Above: Edwin Hubble, the American astronomer whose investigations gave scientists a new sense of the vastness of the Universe. Hubble, using the 100-inch Mount Wilson telescope, discovered that the nearest and largest of the faint misty patches of light that are scattered in the heavens glittered with tiny dots of light. Careful study of these points of light revealed that they were stars of fantastic size and luminosity and were, in fact, separate galaxies multimillions of light-years away.

The Birth of Radio-astronomy

Below: Karl Jansky, the American radio engineer who, in the 1930s, discovered that some radio waves reach the Earth from outer space. He is pictured here with the first equipment ever built especially to receive the signals. The equipment, known as a radio telescope, is composed of an aerial and a receiver. Even with large aerials the signals intercepted from outer space are extremely weak compared with Earth-operated radio waves. The receiver part of the telescope must therefore be a highly sensitive instrument, which in turn means that unwanted radio signals, known as "noise" are picked up, too. The receivers are designed to overcome this interference by greatly amplifying and strengthening the wanted signals.

orbiting an ordinary star that occupied an insignificant position in a lens-shaped collection of millions of stars. Furthermore, our particular cluster—the Milky Way—was only one of many. It is now estimated that there are 3000 million galaxies. The Universe had suddenly expanded.

Proof that the Andromeda nebula (now known as the Andromeda galaxy) consisted of millions of stars and not a cloud of gas and dust came from the detection of Cepheids in its rim. Using these as the newly established measuring device astronomers were able to determine that Andromeda was nearly a million light-years away.

While man was still adjusting to the implications of the new, enlarged Universe, astronomy was preparing to take yet another exciting step forward in its quest for knowledge. Until now all the information about the Sun, the Moon, the planets, and the stars had come to us by one means alone: light. Stand beneath a starry sky and your eyes will collect light from thousands of sources. On cloudy nights you see nothing, of course, and even on clear nights the Earth's atmosphere affects the clarity of vision. In 1931 Karl Jansky, a young American radio engineer, made a discovery that was to change the course of astronomy by opening a new observational window on the Universe. Jansky, working at the Bell Telephone Laboratories, was investigating the static that accompanies all radio reception. During his work he became aware of a faint but steady noise. Having eliminated the usual terrestrial sources for this static he decided it was coming from somewhere beyond Earth's boundaries—outer space.

Above: a contour map of an area of the sky over Cambridge, England. Radio astronomers can now measure and map the radio strengths of various areas of the sky at any given wavelength. Here the contours denote the strengths of radio emissions at a frequency of 160 megacycles (wavelength 1.875 meters). Right: the Parkes radio telescope in Australia. The large radio telescopes are invaluable for pinpointing objects in the farthest reaches of space.

The Sun at first seemed to be the source of the radio signals, but as Jansky continued his observations he found the source of strongest reception was slowly moving away. Over a period of months he followed its course as it crossed the sky. Eventually, it made a circuit of the sky. The young engineer decided that the radio waves were coming from the direction of Sagittarius in the Milky Way, toward our galaxy's center. The importance of Jansky's discovery was largely overlooked at first because ordinary radio receivers could not focus on stellar radio sources with anything like the precision needed to elicit useful information. But an enthusiastic radio ham in Illinois, Grote Reber, took up the discovery and built himself a 30-foot diameter radio telescope with which to study the stars. By 1938 he had found a number of other radio sources, and the new science of radio astronomy was beginning to take form. By 1947, when around 2000 sources of radio waves had been detected in outer space, John C. Bolton, an Australian astronomer, pinpointed the third strongest radio source in the sky to a spot coinciding with the Crab nebula, which is thought to be the remnant of a supernova explosion in our galaxy a thousand years ago. It was the first time a visible

Man Into Space

Below: the Saturn V rocket, the same type that carried the Apollo spacecraft to the Moon. With present rocket technology, space travel is almost certain to be confined to the Solar System. It takes three days for a manned spacecraft to reach the Moon, about five years to reach Jupiter, and 45 years to reach Pluto. Large scale exploration of other planets outside the Solar System, and to other stars and galaxies, does not seem possible for as far into the future as can be seen, because of the enormous distances involved. Even if scientists could design spacecraft capable of traveling at the speed of light—186,282 miles per second—a return journey to the spiral galaxy Andromeda would take almost 4,000,000 years.

object appeared to be responsible for the radio signals. Out of these observations a theory has developed that turbulent gas is responsible for the radio emissions. The theory has been strengthened by subsequent discoveries that Venus, Saturn, and Jupiter, each of which has a turbulent atmosphere, also emit radio signals.

By the early 1960s the Universe had been revealed as far more varied and intricate than even the most daring astronomer had dared to speculate. The early notion of stars of equal size and brightness had been replaced with a cosmic melting pot in which almost anything seemed possible. Stars varied in size, brightness, and behavior. They were grouped in huge island universes. Some blinked at us in a regular manner. Others threw away part of their mass in a great stellar slimming exercises that made them outshine their companions for a short while. Still others departed from the Universe in a gigantic explosion powerful enough to be seen on Earth in daylight. Elsewhere, great clouds of gas were announcing their presence over the space radio network from locations where, for the most part, powerful instruments could detect nothing. What did it all mean? By studying these phenomena, as we shall see, man was able to put forward theories about the origin of stars, galaxies, and the entire Universe.

The decade of the 1960s also saw man's first, hesitant steps into space. Following the success of their earlier, unmanned satellite launches—the first went into orbit in October 1957—the Soviet Union and the United States both put a man into space within a month of each other in 1961. Yuri Gagarin, the USSR's spaceman or "cosmonaut," completed a single orbit of the Earth before landing safely on April 12, and Alan Shepard, the American "astronaut," made a 15-minute suborbital flight on May 5. Man's dream of visiting the stars was suddenly one small step closer to becoming reality. Just eight years after those initial attempts at manned space flight the first human being was walking on the surface of the Moon. American astronaut Neil Armstrong stepped out of lunar module *Eagle* and on to our nearest space neighbor at 02.56 hours GMT on July 21, 1969—a momentous event that was watched by 600 million television viewers on Earth, 232,000 miles away.

Man had speculated about visiting the planets and the stars for centuries but to those television viewers the work of science fiction writers was being overshadowed by the real thing. Will the incredible success of those and subsequent space programs eventually lead man outside the Solar System, so that he can see for himself what the rest of the Universe is like? There is one apparently insurmountable obstacle to that dream: distance. A manned flight to Mars and back using our present rocket technology would take almost two years, because the astronauts would need to wait for the Earth to return to a suitable position before they flew back home. Explorations by man to the limits of our Solar System are not feasible, then, in this century and they will almost certainly have to await the development of new forms of propulsion. As for star travel, that presents even greater problems.

Even if man were to perfect a method of flight that would propel him to the stars at half the speed of light—an incredible 93,141 miles a second—it would still take over eight and a half years just to reach the nearest star-group, Alpha Centauri. If the

Left: Edwin Aldrin photographed on the Moon by fellow astronaut Neil Armstrong on July 21, 1969—man's first day on the Moon. Now that it has been proved possible for man to embark on voyages of exploration into space, scientists can learn much about our Solar System—not only from the instruments the astronauts will be able to set up wherever they land, but, most importantly, from the personal observations of the men themselves.

Below: an artist's impression of a nuclear-powered spacecraft. For long voyages deeper into interplanetary space, the chemically powered engines of the Saturn type, which use liquid oxygen and liquid hydrogen, with helium to start the fuel pumps, will not be practical. To achieve greater amounts of thrust, designers will have to turn to nuclear power and other forms of atomic fuel. Nuclear engines present radiation hazards for the crews, however. In the picture, the crew quarters are placed as far from the nuclear motors as possible and in the shadow of hydrogen tanks, which act as a radiation shield.

space travelers were to spend a year studying the three stars it would be 18 years before the star mission was complete and the astronauts were back on Earth. Another near stellar neighbor is the bright star Sirius—an ideal subject for study because it is a binary star with a dark companion. Even traveling at half the speed of light (which would take the spaceship from the Earth past the Moon in less than three seconds, compared with three days now) the space explorers would be away from Earth for over 35 years. To venture farther afield, particularly to other galaxies, would require huge spaceships and mixed crews who could produce children and train them to continue the exploration as the

The Laser Galleon

Right: an artist's impression of the "space galleon" devised by Philip Norem, the Canadian director of engineering for the Peninsular Research and Development Corporation. It is designed to utilize the laser beam—one of the many advanced kinds of propulsion being proposed for possible future space travel. An array of lasers set up on Earth would drive the space galleon at a speed of 62,500 miles a second toward a planet of the nearest star to the Solar System: Alpha Centauri, 4.3 light-years away. A huge, parachute-shaped sail, made of aluminized plastic and only one micron thick, is attached to the 1000-ton galleon by a cable 20 miles long. The laser beam "wind" would drive the galleon along its course for some $8\frac{1}{2}$ years, after which it would make a wide 180 degree turn. This maneuver, shown in the diagram below, would take between 10 and 30 years to execute, by which time the ship would be approaching Alpha Centauri from the opposite side to the Earth and it could use the laser wind as a brake. The ship would then coast down onto its target planet, taking another $8\frac{1}{2}$ years. The astronauts, by then much older, would set up a laser array on the planet as a brake. Subsequent journeys would take 10 years instead of the 60 or so taken by the pioneers.

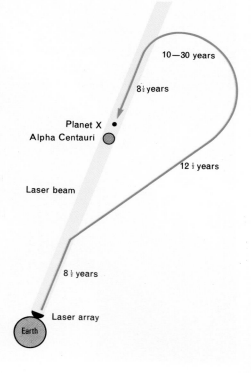

Relativity and the Universe

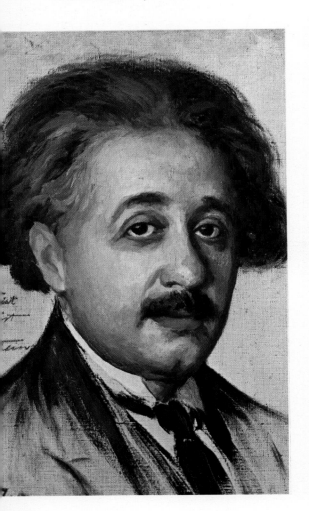

Below: Albert Einstein, the German-born mathematician and physicist. His theories of relativity changed scientific concepts of both space and time. He demonstrated that time and space are not absolute but relative to the observer and also that the rate of time is affected by mass.

more elderly members of the space mission died.

Man's thirst for knowledge will almost certainly inspire him to make such space explorations in the next few hundred years and when he does he may, after all, find that the distances are not as daunting as they now appear. The reason for this is contained in the now celebrated equations of Albert Einstein, a German-born mathematical-physicist. His Special Theory of Relativity, which was published in 1905 suggests, among other things, that as speed increases time slows down. This effect is not measurable at the speeds with which we are presently familiar, but if we were able to travel close to the speed of light it would have a very real effect. The clocks on board the speeding spaceship would run more slowly and the atomic and biological processes of everything on the craft would slow down correspondingly. So a trip to the stars would seem to take months instead of years. A man returning from a journey made at such speeds would find that his twin brother left behind on Earth was much older than himself, which is why this effect is known as either the clock paradox or the twin paradox. Even more incredible is that such a space traveler could return to find his own grandchildren older than he is. A journey to the stars, it seems, would provide the source of eternal youth, for if the spaceship attained the speed of light, time would stand still.

How can this possibly happen? The reason is that Einstein, building on work done between 1899 to 1904 by Frenchman Jules Poincaré and Dutch physicist Henrik Lorentz, conceived a four-dimensional Universe in which time was one of the dimensions. Einstein's statement of Special Relativity was undoubtedly a major contribution that led to the new concept of "space-time" as a continuum in which it was impossible to isolate one from the other.

Einstein's view of the Universe is mathematical and although those aspects of his theory that can be tested have been confirmed experimentally, it still does not make it any easier for the layman to grasp. For centuries the popular idea of the Universe modeled on Newton's simple and mechanical laws has sufficed. By giving time a quality it does not seem to possess on Earth Einstein's theories, however accurate, make the Universe more mystical and difficult to comprehend. Fortunately, the effects of theory of relativity occur only at speeds approaching that of light. At slower speeds Newtonian and Einsteinian laws are virtually the same.

Time and space seem to us on Earth to be totally different concepts, so it is difficult to understand how they can be linked. As far as space is concerned, we can go up and down, side to side, or backward and forward. In theory, at least, we can go where we choose at any speed and for any distance. We can even stop and change direction. Time, on Earth, offers no choice. It marches on relentlessly: the seconds, minutes, hours, days, months, and years tick away with precision. We cannot make time stand still, nor can we travel back in time. There is only one direction in which we can travel and one speed at which we can progress. And this, commonsense tells us, must be the case everywhere else. So how can space and time be linked?

We know, for instance, that it is a simple matter to pin-point the exact position of an object in space by making a number of measurements, such as the distance between us and it, and between it and another object, and the angles between all three. But if we give those measurements to someone and ask him to locate the object he may not be able to do so. It may have moved. Those measurements relate to a moment in time and we need to measure time in relation to a moving object if we are to gain knowledge of all its properties. And that measurement of time depends on our viewpoint in space.

A spaceship may appear to be rushing toward us at an incredible speed but to its occupants the spaceship would seem to be at rest and it would be Earth that was racing toward them. Who is right? Motion in space is relative to the movement of other objects and the answer to such a question depends on the vantage point from which you view events. If you were on a stationary train and you bounced a ball gravity would make it fall and rise vertically, bouncing on the same spot. It would appear the same whether you were inside the train or looking in from outside. Later, when the train is traveling at 100 mph, if you were to start bouncing the ball again the same thing would happen. It would rise and fall on the same spot and everyone on the train would confirm that no change had taken place. But to an observer standing and watching the train a very different picture would present itself. He would see the ball dropping and rising at an angle because in the time it takes the ball to fall from your hand to the floor the speeding train will have moved forward several feet and the pull of gravity will make the ball return at an angle.

Now supposing two 300-foot spaceships were to pass each other in opposite directions at 162,000 miles a second and they were each capable of measuring the length of the other as they flashed by, they would each determine the other to be only 150 feet long. The occupants of each ship, however, would be aware only of the difference in the other craft. If they were to measure their own craft's proportions at that very moment it would still be 300 feet long. Similarly, if they could see the clocks on each other's spaceship they would believe the other's to be running at half the normal rate, while theirs was normal. If the two spaceships were to slow down and stop next to each other they would find that they were then both back to normal size. But what about their clocks? Einstein's Special Theory of Relativity did not allow such a comparison to be made because it concerned only subjects that traveled at a constant speed and in the same direction forever. In 1916, however, Einstein produced his General Theory of Relativity, which included situations in which objects changed speeds and directions. According to this, if two spaceships pass each other at the same speed, then one turns around and accelerates in order to catch up with the other, and the two came to rest and compared clocks, those timepieces on the craft that had increased its speed would show that time had passed at a slower rate than on the other spaceship. Time, then, moves more slowly for objects accelerating in space relative to the rest of the Universe, but time on board the object traveling at this high speed would appear to be normal.

The General Theory of Relativity also gives us a new way of looking at gravity. Instead of it being a mysterious power exerted by the mass of a body, Einstein argued that it was a property of

Above: a cartoon by George Strube, the British artist, made in 1930. Einstein's Theory of Relativity was generally considered to be an extremely difficult concept to understand. This cartoon of Einstein talking to the "ordinary man" reveals a simplified way of looking at his theory of relativity.

Above: a solar eclipse, with the Moon's disk all but hiding the Sun. The edge of the Moon's disk looks irregular because the Sun's rays bend around it rather than meet it edge on. In his General Theory of Relativity, Einstein declared that any mass tends to distort or "bend" space. Like space, light, too, is distorted by a mass, as modern photographs of an eclipse have proved.

space and that mass warps space and time. This can be demonstrated with a physical model: Imagine space-time as a two-dimensional stretched rubber sheet onto which a heavy sphere, representing the Sun for example, has been placed. It will depress a portion of the sheet. Now roll a smaller sphere across the sheet so that its path takes it into the depression around the first sphere. Depending on its size and speed it will either be captured by the larger sphere and be unable to escape from the depression, or the curved space will deflect it from its original course. Planets, such as Earth, are captured in the curved space caused by the Sun's mass but centrifugal force counteracts this, keeping us from being pulled into the Sun. According to Einstein even a ray of light passing close to the Sun's surface would be deflected by this space/time curvature and astronomical experiments during solar eclipses have found this to be so.

The Universe as seen through Einstein's eyes requires a familiarity with mathematics for a total understanding. Without that it is difficult for most people to visualize all the consequences of relativity because they are alien to what we are accustomed to regard as normal. Experimentally, however, there is every reason to assume that Einstein is right even if, in time, new theories evolve that will take us even nearer to understanding the workings of the Universe.

We should, perhaps, hope that Relativity will one day prove to be only part of the story, and that the limits it contains will be seen to be non-existent, otherwise star travel will be almost impossible. Einstein's theories insist that although time can slow down it can never run backward. They also say that 186,200 miles a second is the maximum speed that can be achieved by any accelerating object below the speed of light. Time travel—the science fiction writer's trump card—would seem to be an impossibility—now and for ever—unless the light and time barriers are found to be no more real than the sound "barrier" that once seemed to impose a limit on aviation.

The Universe, then, is far from that envisaged by the early civilizations: ruled by gods and goddesses who manipulated the Sun, Moon, planets, comets, meteors, and stars, in order to keep man in order. But scientists in other fields have made discoveries that have caused some to reconsider the claims of the early astrologers. Shorn of the mumbo jumbo of occult jargon, and deprived of the mystique that enshrouds the subject, some basic principles of astrology seem to have been confirmed in a quite remarkable way. The conclusion that many draw from this research is that man, and all life on this planet, is subtly influenced by a vast range of cosmic influences.

We know from our study of the Universe that unseen forces not only control atoms and bind solar systems but can also cement huge galaxies together. Why, then, should it be thought absurd that similar forces can affect the lives of men? The mystery is how this could be achieved. Some scientists believe that we are fast approaching the time when we will understand all the mechanics of the Universe. If that is so, then perhaps man will start a new investigation of the more subtle influences that may be at work in every living cell upon our planet—and upon every other inhabited world in the vast, mysterious Universe.

Gravity in Space

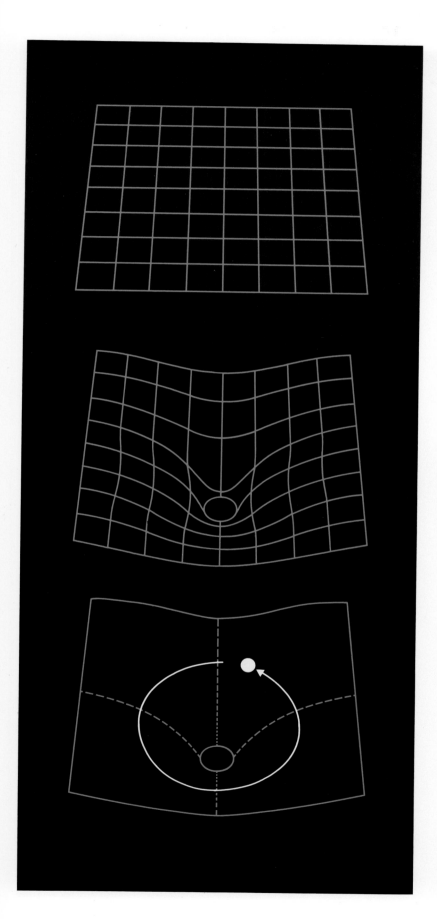

Left: a diagrammatic representation of Einstein's four-dimensional world of the relativity theory. To show what happens when moving material bodies influence each other, the analogy of the rubber sheet became popular. It illustrated the theory of a three-dimensional space to which was added a fourth dimension, time—measured as distance traveled by a given interval of light. The first diagram shows space represented as a flat rubber sheet marked with parallel lines.

Center left: the flat rubber sheet representing space is distorted by the weight of a star's mass. The area around the mass is now curved.

Bottom left: because of the distortion of space around a mass—such as a star—satellite planets are held in orbit rather like racing cars circling on a banked racetrack.

Chapter 3
Islands and Holes in Space

What were the signals from outer space that radio astronomers were receiving so regularly in the 1960s? Were other beings trying to make contact? For hundreds of years new and fascinating puzzles have arisen in the Universe for astronomers and astrophysicists to solve. Some refuse to be solved easily. For example, it was not until 200 years after its discovery that a "fuzzy star" in Andromeda was found to be a gigantic island Universe millions of light-years away. How long will we have to wait before we solve the puzzle of the most awe-inspiring object in the entire Universe: the black hole?

The bright city lights of Los Angeles were blacked out in 1942 during World War II. With their glare no longer obscuring the night sky, Walter Baade, the German-born astronomer, seized the opportunity he needed. He had the giant 100-inch Mount Wilson telescope, perched high above the city, trained onto a tiny patch of light in the Andromeda galaxy—the huge galaxy so far away that only a few stars could be seen with clarity. Taking advantage of the improved visibility that the wartime conditions had provided, Baade was able to see not only the stars in the spiral arms of the huge galaxy but also some at its center. He realized immediately that there were important differences between the two. Those near the center of this great "island" in space were reddish, whereas those on the outskirts were bluish. His observations also revealed that the brightest of the blue stars were many times more luminous than the red giants at the galactic center, and that while the outskirts of Andromeda were littered with dust, its interior was free of this space debris.

Baade classified the bluish stars as Population I and the red stars at the center as Population II. He continued his observations with other galaxies and soon had the use of the new 200-inch Hale telescope at Mount Palomar, which gave a far clearer picture of individual stars in many of the galaxies. Much investigatory work had already been done some years earlier on the galaxies by Edwin Hubble, the United States astronomer at Mount Wilson. He was the man who had first detected individual

Opposite: the Andromeda galaxy. It has a special interest for astronomers because it is closely similar in size and mass to the Milky Way, which contains our Sun and the Solar System. Andromeda can be seen from the Earth with the naked eye as a faint yellowish blur of light. The yellowish blur is only the bright central part of the galaxy—the part, in fact, made up of old stars with low surface temperatures. The outer regions contain blue stars, newly formed from gas and dust.

How the Galaxies were Formed

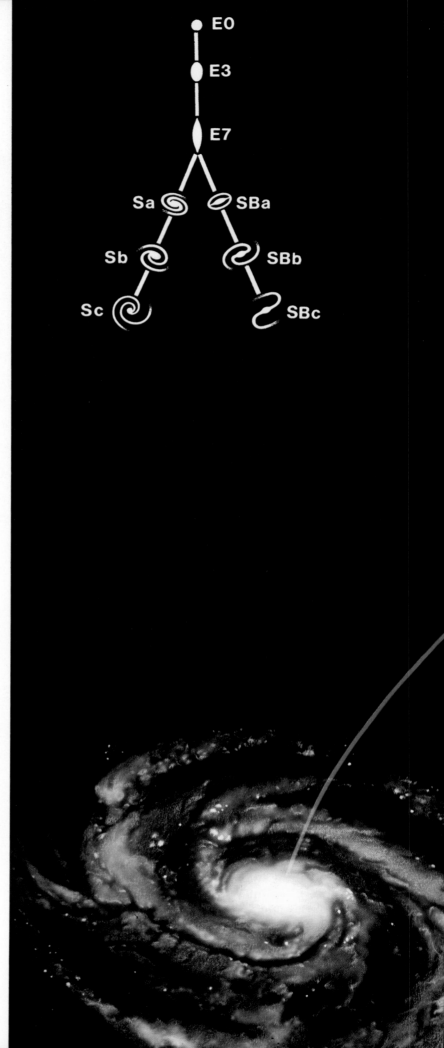

Right: how galaxies are classified. The illustrations on the far right show how the galaxies appear in the sky. The key, top left, shows the galaxies in diagrammatic form, labeled according to the method laid down by the famous American astronomer Edwin Hubble. In the handle of the fork are the ellipticals, denoted *E*. Ellipticals range in shape from globelike to the purely elliptical. The roundest ones are referred to as *E0* galaxies, and then the numbers go down in sequence to more oval ones, the most eliptical being *E7* galaxies. The prongs of the fork are divided into spirals (*S*) and barred spirals (*SB*). On the left fork, the degrees of tightness of the spirals are lettered *a*, *b*, and *c* so that we can designate a galaxy from this type as *Sa*, *Sb*, or *Sc*. The same lettered designation is applied to barred spirals also, so that they are labeled SBa, SBb, or Sbc.

The normal and barred spiral galaxies differ considerably in size, the smaller ones are around 30,000 light-years in diameter and the larger ones appear to have a diameter of about 120,000 light-years. An average sized galaxy contains enough material to create about 2000 million stars the size of our Sun. Most of this material is contained in the stars that make up the galaxy, but there is also a large amount of material scattered as gas and dust. The elliptical galaxies, on the other hand, are largely free of gas and dust and as it is from this gas and dust that new, bright stars are born, they are usually dimmer than the spirals. One theory, held by many astronomers, is that different shaped galaxies are really different stages of development. It is believed that galaxies start life as irregular-shaped galaxies, such as the Magellanic Clouds, and gradually evolve through spirals, which perhaps "wind themselves up" into ellipticals and so on into sphericals.

The Galaxy Types

Below and opposite: photographs of nine galaxies taken with the 60-inch telescope at Mount Wilson Observatory. They are labeled according to Edwin Hubble's system of classification. Starting at *E0*, the elipticals become increasingly flattened through *E5* and *E7*. Spirals are more easily identifiable where the galaxy nucleus is small. With a few exceptions, all the galaxies observed by Hubble fitted into his classification.

E0   **E5**   **E7**

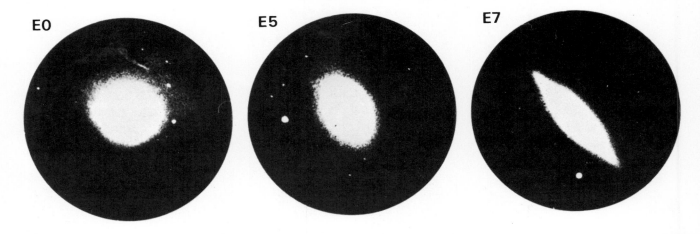

Right: a galaxy in the northern constellation Triangulum (the Triangle), which is between the constellations Pisces and Perseus. Hubble's classification puts it into the *Sc* category of spiral galaxies.

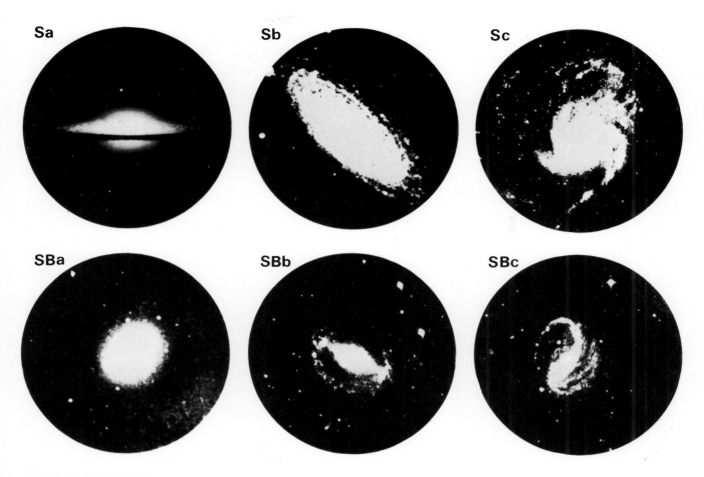

stars in Andromeda, thus establishing that it was a huge collection of stars very far away and not a nearby cloud of dust and gas. Through his work astronomers realized that there were three main types of galaxy: spiral, ellipsoidal, and irregular. But within each main type there were differences, such as the degree of flatness, so Hubble gave each a classification. An ellipsoid was shown as E, and E1 through E7 represented degrees of flatness. Spiral galaxies appeared to be of two kinds: the ordinary, which were denoted with the letter S, and those that had a "bar" running through them, which were given the symbol SB. He then classified each according to the arm formation: a, b, or c. Sa, for example, was a normal spiral with tightly wrapped arms while SBc was a barred spiral with much looser structures. Our own galaxy was of the Sb type, very similar to Andromeda. It seemed to Hubble that each of these types represented an evolutionary stage in the life of a galaxy, and his theory was that galaxies began life as spheroids and, as each contracted and grew more compact its rotation increased causing it to flatten and throw off arms at the outskirts.

Baade's more detailed observations of the stellar composition of the galaxies reversed this view. The bluish stars were young with a high metal content while the red ones were old and of low metal content. Elliptical galaxies were found to be made up of mostly the red stars, and dense groups (*clusters*) of stars in any type of galaxy proved to be of the same type. The new picture of galactic evolution that emerged was that the spiral arms that are

Right: the great spiral galaxy of Andromeda and its elliptical companion galaxies M32, below right of the central region, and NGC 205, above right.

Below: detail from part of one of Andromeda's spiral arms shows the bright bluish stars and giant stars that make up the arms of this type of galaxy. They are young O- and B-type stars categorized as Population I stars.

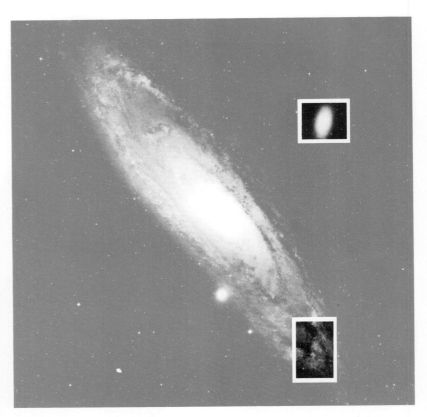

Below: a detailed view of Andromeda's companion galaxy NGC 205. In this typical elliptical galaxy Population II stars predominate—that is, A, G, F, K, and M-type older stars. They are also found in the nuclei of spiral galaxies.

common to so many galaxies are a short-lived phenomenon and that the tendency is for galaxies to lose them as they wrap themselves closer to the nucleus—and form into ellipsoids. But if that is the case, a mystery remains. Why do so many galaxies have spiral arms if they are just a short-lived phase?

The stars that predominate in the spiral arms are the blue ones and there is thought to be only two percent of such stars in the Universe. Our own Sun is of this type and we can deduce from that fact alone that we are in the spiral arm of a galaxy among great clouds of dust—a fact that is confirmed by observations of a totally different kind.

During his study of galaxies Baade turned his attention to the Cepheids—the variable stars that had enabled astronomers some time before to measure the vast distances between us and the more remote objects in the Universe. He soon made a discovery of tremendous importance. Whereas the Cepheids at the center of the galaxies had the correct brightness according to the period-luminosity curve established by Henrietta Leavitt, the Cepheids in the blue stars were four or five times as luminous. So if, unwittingly, a distance had been measured using a Cepheid in the spiral arm of a galaxy it would be totally wrong. A revised distance scale was calculated and Andromeda was proved to be 500,000 light-years away. The other galaxies shifted farther from us, too, for the effect of this discovery was to more than double the size of the known Universe.

Baade's observations also cleared up a number of puzzling aspects about the galaxies. In comparison to our own, most others had previously seemed to be smaller and their globular

Our Neighbor Galaxies

Left: a plan of the Local Group, the cluster of galaxies containing our own Milky Way. There is still much disagreement among astronomers about which galaxies to include. In this plan, the Milky Way is in the center and Andromeda, our "mirror image," is seen to the right labeled M31.

clusters less bright. Now, with the new evidence about their distance from us, it was clear that ours was not, after all, an exceptional galaxy. In fact, Andromeda is larger than our galaxy though we remain a giant in comparison to many. The most massive known galaxy is M87, which could contain as many as three million million stars. In time it also became apparent that just as planets revolve around stars and stars are drawn together into huge galaxies, so the galaxies themselves seem to be attracted to each other and move through space in groups. Our own galaxy, the Milky Way, is part of the Local Group of around 20 members, including the Megallanic Clouds, Andromeda, and three Andromeda "satellite" galaxies. Two members were not discovered until 1971 because they were obscured by space dust, so there may be still others that we have not yet detected.

Depending on the brightness of objects in space, they are referred to as *dwarfs* (low luminosity) and *giants* (high luminosity). Most of the members of our Local Group are dwarfs (we and Andromeda are exceptions) and, throughout the Universe, dwarf galaxies outnumber the giants. The average distance between galaxies is thought to be about 20 times their diameter. Some clusters, however, are more closely packed than that, and some are truly immense. The Virgo cluster contains over 1000 galaxies and the cluster in Coma Berenices consists of 10,000 galaxies at a distance from us of 400,000,000 light-years. Measurements made in 1965, incidentally, suggest that our galactic center is smaller than we once thought, in which case our spiral arms are more prominent and widely spaced than we had visualized. If this is so our galaxy would look more like the Whirlpool

Below: one method of determining whether objects in the Universe are approaching, receding, or remaining stationary in relation to the Earth's position. This form of measurement was first discovered in relation to sound by Christian Doppler, an Austrian physicist, in 1842. When a source of sound—or light— is in motion the sound or light appears to undergo a change in frequency to a stationary observer. The frequency will be high or low depending on whether the source is approaching or receding. In sound, the whistle of an approaching train is pitched high; as the train passes, the whistle has a lower pitch. Moving light shows itself in a shortening of wavelength in an approaching object, a lengthening of wavelength in a receding one. By observing light from celestial objects through a prism, light displaced toward the red end of the spectrum means that the source is receding; toward the blue end, that the source is approaching. In the diagram below *A* illustrates a light source from an object traveling at the same speed as Earth; the spectral lines are in the normal position. *B*, a light source approaching Earth; the spectral lines shift to the blue end of the spectrum. *C*, a light source receding from Earth; the spectral lines shift to the red end of the spectrum.

galaxy (an Sc classification) than Andromeda.

As well as giving us an insight into the astonishing scale of the Universe, study of the galaxies has also enabled scientists to argue with confidence about the way the Universe was born. This has come about through what is known as the *Doppler effect*, named after Christian Johann Doppler, an Austrian physicist who in 1842 first explained a strange phenomenon associated with sound. If you are standing on a railroad station and a train races through blowing its whistle, the sound alters very suddenly as it passes you. But someone on the train would hear the sound of the whistle at the same pitch all the time. The reason, Doppler explained, is that sound travels in waves causing compression of the air. The distance between each compression is the wavelength of the sound, and the longer the wavelength the deeper the sound. If we listen to the whistle while the train is stationary we hear a constant pitch as the compressions radiate out at regular intervals. But when the train begins moving toward us, still sounding its whistle, each successive region of compression will reach us that much sooner than when it is stationary. This, in effect, is shortening the wavelength as far as we are concerned and so the pitch becomes higher. But the moment the train races past us the opposite happens. It now takes each region of compression longer to reach us as the source of the noise travels away. This "stretches" the normal whistle wavelength into one that is longer, and so the sound suddenly drops in pitch. The faster the train is traveling, the higher the whistle's pitch on its approach to you and the lower it is when it speeds away. So, by knowing the

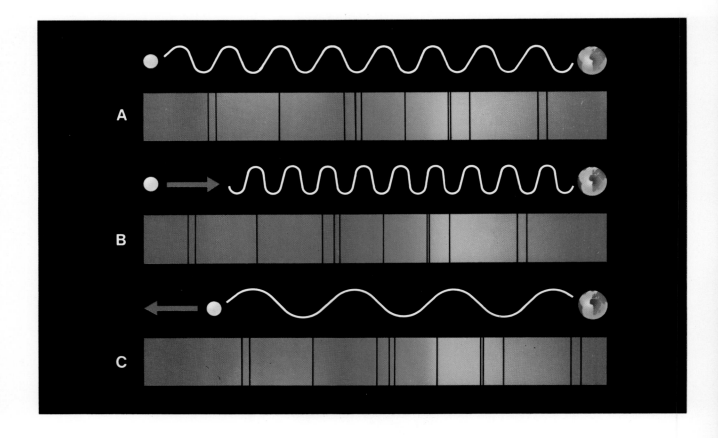

train's normal whistle pitch it is possible to calculate, just from its sound, whether it is coming toward you or going away and at what velocity it is traveling.

The same is true of light, since it too is a wave form radiating from a source. We detect a difference of wavelength in light by the color it emits. The longest wavelength visible to us is red and the shortest is violet. So, as Doppler himself commented, it should be possible to detect whether an object is moving toward us or away from us, and its speed, by studying the light it gives out. The way to do this, as far as stars and other objects in the sky are concerned, is to pass their light through a prism and examine their spectra. This well-known experiment was used to detect the existence or absence of elements in different substances and had already been used to study the make-up of stars. Armand Hippolyte Louis Fizeau, a 19th-century French physicist, pointed out that the Doppler effect should cause the spectral lines related to the elements to shift if the stellar object was moving. A move toward the red end of the spectrum (the longer wavelengths) by the spectral lines would indicate that the star was moving away from us, while a shift to the violet end meant it was coming toward us. The amount of shift would indicate velocity. This is usually known as the *Doppler-Fizeau effect*. But because light travels at such fast speeds the shifts are not easily detected, and it took until 1868 for William Huggins, a British astronomer, to prove in this way that Sirius was moving away from the Sun.

It was not until the early 1900s, however, that these luminous stellar objects were studied to determine their velocity. They were

The Doppler Effect

Below: a cluster of galaxies in Corona Borealis. By observing the spectrum lines of this constellation, which is over 1000 million light-years away, astronomers calculate that it is receding from us at an incredible 13,400 miles (21,600 km) every second. Scientists have found that the farther the galaxies are away from Earth, the faster their speeds of recession. Does this mean that there is a limit to how much of the Universe we shall ever be able to observe? Using the speed of recession table, the farthest galaxies must eventually reach the speed of light—so that those galaxies will for ever remain invisible. Because Corona Borealis is so far away from us, the form of the individual galaxies is hard to make out. A good way to distinguish them here is to eliminate the small round spots of light and the objects with spikes, which are stars, and the remaining objects are galaxies.

Radio Galaxies

Right: the radio source, Cygnus A in the constellation Cygnus. Its powerful radio emissions were first detected in 1946 and five years later were pinpointed as a position 550 million light-years away. When finally photographed with the 200-inch Mount Palomar telescope, observers saw that it was a galaxy that appeared to be undergoing a cataclysmic process. Because Cygnus A is so far away, and yet is the second most powerful radio source known, astronomers have given much thought to what could be causing such intense radiation, and the true nature of the curious double structure of the galaxy.

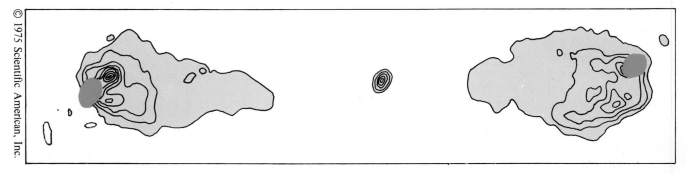

Above: contour map of the strong radio source in Cygnus A. The blue spots at either end represent areas of greatest intensity of radio-wave emission. Maps such as this have enabled radio astronomers to build up a picture of Cygnus A and similar radio galaxies. From the parent galaxy (the structure in the center of the map) jets of intense radio waves are ejected from both ends—in this case, to a distance of some 500,000 light-years apart. It is these main lobes of radiation that give the source its double-galaxy appearance. There is also a weaker source of radio waves from the central parent galaxy.

all referred to as nebulae, though those that were later seen to be made up of stars were renamed galaxies. By 1917, out of 15 "nebulae" examined by Vesto Slipher, an American astronomer, 13 were found to be receding from us and two were approaching. One of those approaching was Andromeda. What was particularly disturbing to astronomers at the time was the velocity at which these bodies were receding: 400 miles a second. When, a few years later, Edwin Hubble showed that these objects were great star clusters very like our own galaxy and were at great distances from us, their speeds were more acceptable, though still quite astonishing. Further study showed that nearly all the galaxies were receding from us at tremendous speeds. We know now that the farther a galaxy is from us, the greater its speed of recession.

With the development of radio astronomy in the 1930s science was provided with a new tool with which to tackle some of the Universe's many mysteries. Astronomers were no longer restricted to observing only those objects that were large enough or near enough to be seen according to the light they emitted. Many stars too small or too distant to be seen could be detected through the sounds—as radio waves—they gave out. By 1948 the Australian radio astronomer pioneer John C. Bolton had narrowed down a puzzling source of radio waves, Cygnus A, which

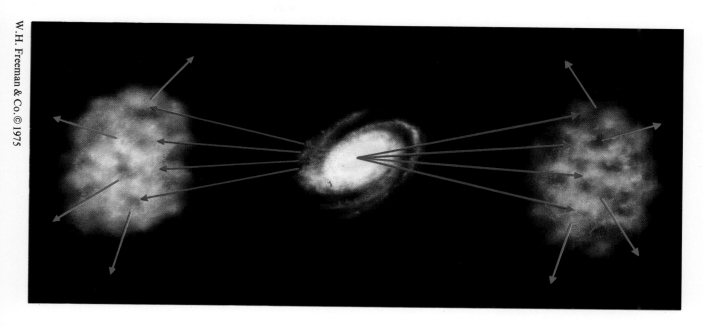

was sufficiently sharp for him to name a "radio star." Three years later he was able to indicate with even more precision the spot in the sky from which this powerful microwave emission was coming, and Walter Baade studied the area with the 200-inch Mount Palomar telescope. His search was rewarded when he detected an oddly shaped galaxy which, on closer inspection, looked like two galaxies colliding with their nuclei close together. It was suggested that the other mysterious sources of radio waves in the heavens might also be from strangely shaped galaxies. Very soon 100 "radio galaxies" were discovered. Many are very peculiar in appearance, but the belief that the intense radio emissions were caused by colliding galaxies was short-lived. Argument upon argument was put forward to show why the microwave emissions could not be the product of such a cataclysmic disaster. More likely, it seemed, was the possibility that radio galaxies with microwave emissions 100 times stronger than ordinary ones were single, exploding galaxies.

Comparative study of different radio galaxies seems to show them at various stages: those with weak microwave emission from near their centers seem to indicate an early stage whereas those with stronger radio sources emanating from far greater areas of the galaxy indicate that the explosion is in an advanced stage. But what causes a galaxy to explode? One explanation is that when numerous red stars at about the same time reach the stage of spectacular luminosity that usually precedes an explosion —a supernova stage, sometimes as much as 200 million times the brightness of our Sun—one exploding supernova will trigger off the others causing the wholesale destruction of the galactic center. But to account for the microwave emission of Cygnus A we would have to assume that in excess of 10,000 million stars like the Sun were involved if hydrogen fusion is responsible for the radio waves. A more likely suggestion comes from Fred Hoyle, the British astronomer, who argues that in a very crowded galactic center the gravitational attraction between the stars could overcome the forces that normally keep

Above: diagram showing one theory of how radio galaxies such as Cygnus A produce a double pattern of emission. Explosions of great violence triggered off in the very heart of the spiral galaxy (center of diagram), cause streams of fast moving particles (blue) to be shot out in opposite directions at a speed approaching that of light. If the streams of particles are checked by gas clouds, the magnetic field that always exists in such clouds trap the particles and divert them into rapid motion around the magnetic field. This causes the particles to emit radio waves (red).

The Mystery of the Quasar

Below: Martin Ryle, the British Astronomer Royal. Working with colleagues at Cambridge, England, he has defined various categories of radio emissions—called brightness—from strong to weak, and a scale of distances.

Right: the quasar 3C273. The bright jet, outlined in black in the photograph, is a strong radio source. Radio astronomy has played an important role in clarifying and adding much new information on many already known objects in the Universe. With the pinpointing in 1963 of 3C273, however, radio astronomy had discovered the quasar—a completely new category of celestial object. The starlike quasars (many more have been discovered since 3C273) are named for quasi-stellar objects. They are among the most puzzling phenomena in the whole of the Universe.

them apart. As they were drawn closer together the gravitational field would grow stronger, attracting even more stars until perhaps 100 million stars rush to the center to form a single super-star. Such an event would explain the high microwave emission, but for the time being the true nature of this tremendous radio energy remains a matter of great speculation—a still unresolved mystery.

In addition to the radio galaxies there still existed a number of radio sources in the heavens that seemed unusually small. Martin Ryle, an English astronomer, working with a group of other astronomers compiled a list of these sources—known as the Third Cambridge Catalog of Radio Stars—and gave each radio wave source a number, preceded by 3C for the name of the catalog. A team of Australian radio astronomers were among those trying to identify the source of the emissions and they devised an impressive way of isolating the tiny area of the sky from which one of the sources, 3C-273, was coming. Realizing that the Moon was going to pass between this source and the Earth they decided to track it and record the very moment when the edge of the Moon eclipsed it. The astronomers, feeling their study could be important, went to extraordinary lengths to ensure the success of the experiment.

They sawed several tons of metal off the telescope to allow observation at a lower angle of elevation than normal. Then, for hours before the lunar eclipse in 1962, local radio stations broadcast appeals that no one should switch on a radio transmitter during the period of observation. Patrols were set up around the observatory to stop the movement of vehicles. Following the observations, duplicate records were made and taken from the observatory in Canberra to Sydney, 150 miles away, by two members of the team on separate planes to ensure that at least one arrived safely for analysis. All the effort proved worthwhile. They pinpointed the exact position of radio source 3C-273, which proved to be a double source of microwave emission, they passed

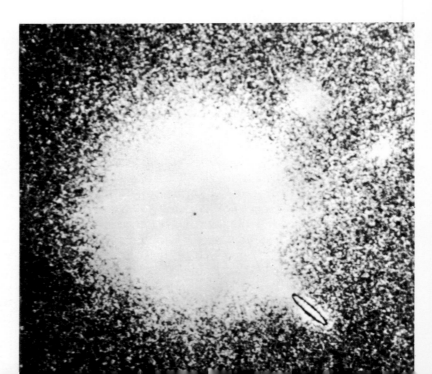

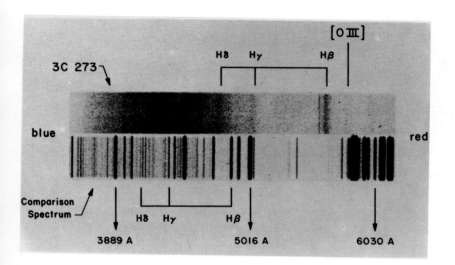

Left: the spectrum produced by light from quasar 3C273 (top) compared with a laboratory spectrum for measurement. The hydrogen emission lines (Hδ, Hγ, and Hβ) marked on both spectra show the enormous red shift of the quasar. According to the spectrum, astrophysicists have worked out that the quasar is 1000–2000 million light-years away.

Below: radio emission from the quasar 3C273 as recorded with the 250-foot radio telescope at Jodrell Bank, in Britain. The intensity of such emissions, together with its apparent enormous distance from Earth, seems to indicate that 3C273 is as bright as 200 galaxies combined.

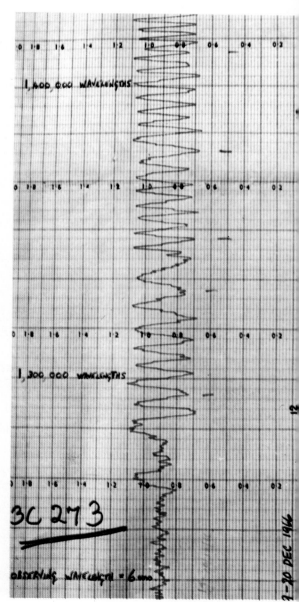

the information to optical astronomers at the Hale Observatories in California. Checks showed that an object was known to exist in exactly the position of one of the components but it had always been regarded as an unimportant star in our own galaxy. The star has a jet coming from it that ends at the point where the second radio source had been located. Spectral analysis of the light from this object produced four emission lines that did not correspond with any known spectral lines. One possible explanation was that the emission lines could have shifted (as they do with galaxies that are receding from us) and further study showed that if this was so, 3C-273 must be traveling away from us at nearly one-sixth the speed of light. These objects have since been named quasi-stellar (starlike) sources, and shortened to *quasar*. Quasars have become the most puzzling astronomical objects of recent times.

What, then, are quasars? There are two explanations, both of which give them very unusual properties. If the spectral lines are the result of a red shift caused by extremely rapid recession from us then the quasars must be unlike any stars that we have seen in our part of the Universe. They would have to be billions of light years away. By the end of the 1960s 150 quasars had been found and the spectra of more than 100 were studied. In every case they showed a large red shift, which meant the objects were receding. And yet, if they are so far away (some are estimated to be at a distance of 9000 million light-years) then they are between 30 and a 100 times as luminous as an entire galaxy. If they were massive galaxies, containing as many as 100 times more stars than an ordinary one and five times as large, which would explain their great luminosity, then even at their great distances our most powerful optical telescopes should be able to see them as recognizable oval misty blotches. Instead, they are just small starlike points of light.

In 1963 the quasars were found to have a variable output of radio energy, which is a very strong argument in favor of believing they are small bodies. In which case, if they are small, distant, and amazingly bright, they must be using up energy at an astonishing rate, giving them a life expectancy of just a million years or so—a very short time by cosmic standards.

Evidence to support this theory came in 1965 with the discovery of what appeared to be aged quasars. They looked like common bluish stars but they had the same great red shifts as quasars. They were as distant and as luminous as small quasars but they did not radiate microwaves. They were given the name "blue stellar objects"—since abbreviated to BSOs—and they are now known to be more common than quasars with perhaps as many as 100,000 within the reach of our telescopes. By the early part of 1977 astronomers had discovered a quasar racing from us at 98 percent the speed of light: an incredible object traveling at an incredible speed and—with its companions—the subject, for the time being, of intense speculation and controversy.

While many of the quasar discoveries were being made, a team of British astronomers at Cambridge were building a special radio telescope capable of detecting and studying very short bursts of microwave energy. It had been noted that some sources were capable of changing their intensity of radio emission very rapidly but a specially designed telescope was needed to study the phenomenon properly. Very soon after it came into operation the team, led by Anthony Hewish, recorded extremely short bursts that lasted precisely one thirtieth of a second, and with an interval between each burst of radio energy of precisely 1.33730109 seconds. No visible counterpart could be found for these strange signals which, at first, seemed awesomely like an attempted communication from another form of life. But the signals were too monotonous to be intelligent signals: there had to be another cause. Because of their remarkable nature they were named pulsating stars, which was shortened to *pulsars*. Hewish had found four by the time he announced his discovery in February 1968. Other astronomers soon found more and a total of 40 were located within two years. Whereas quasars are extremely distant, pulsars are, generally, in our own galaxy (they might well be present in others too, but they would be too faint to detect with our present instruments). And unlike quasars it was possible to offer a convicing explanation of these strange and fascinating objects very soon after their discovery.

In 1934 Fritz Zwicky, a Swiss-American astronomer, had suggested that under certain conditions the subatomic particles in a star might combine to become neutrons, which would be com-

Pulsars – Signals From Outer Space?

Opposite: part of the 3-mile-long radio telescope near Cambridge, England. In all, there are eight dish aerials, four of which can be moved along the old railroad track over which the assembly is built.

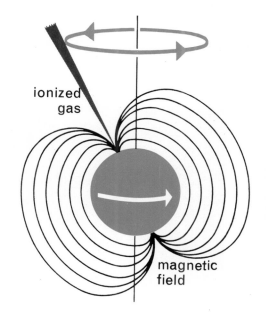

Above: there is much speculation about how exactly a pulsar—a pulsating star—works. One theory is that ionized gas jets up from the magnetic field lines, emitting radio waves. As the star rotates it beams out a radio pulse like some distant radio-wave "lighthouse" in space.

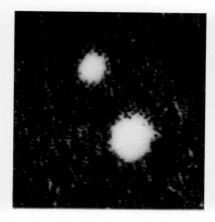

Left: the pulsar in the Crab nebula. It gives out pulses of radiation as visible light, x-rays, and radio-wave activity. The photograph far left shows the pulsar "on," left shows it "off."

The Evolution of a Star

Below: the stages astronomers believe a star, such as our own Sun, passes through. The rotating cloud of gas and dust, extreme left, condenses to form a globe and perhaps the beginning of a planetary system (second picture). The globe condenses further and nuclear reactions transform it into a true star (third picture). Over millions of years it expands to become a red giant (pictures four and five). It later becomes a pulsar (six), and finally a "dead" white dwarf.

pressed together until they made contact with each other. Under such conditions a very massive star would be compressed into an extremely dense neutron star that, though no more than perhaps 10 miles in diameter, would still have the mass of its original form. It was a theoretical model that was resurrected in an attempt to explain certain cosmic X-ray phenomena in the early 1960s, but had been buried when it did not seem to fit the facts. It was Thomas Gold, a British astronomer, who suggested it might explain the pulsars, since to pulsate at such brief intervals the object in question had to be rotating at tremendous speeds, and for that to occur it had to be small and very hot, or have great gravitational fields. A neutron star, said Gold, would be small enough to revolve on its axis in four secor   less. It would also be leaking energy at its magnetic poles (which would appear to us as short bursts of microwave emission each time a pole pointed in our direction) and this would cause it to slow up gradually.

In November 1968 astronomers at Green Bank, West Virginia, found a pulsar in the Crab nebula—the center of a great deal of extraordinary cosmic activity. The Crab pulsar was pulsating 30 times a second—the most rapid pulsar known. Careful study showed that it was, as Gold had predicted, slowing down by a rate of 36.48 thousand millionths of a second each day. The same phenomenon was found with other pulsars and the neutron star theory became accepted as fact. Since the microwave emissions

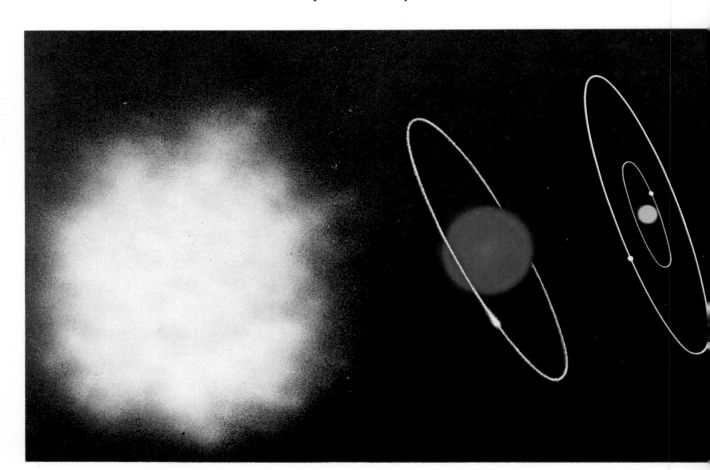

were caused by electrons escaping from the condensed star it was likely that the pulsations could be detected visibly, too. And so it was, for in January 1969, a dim star within the Crab nebula was found that did flash on and off precisely in time with the radio signals. So the Crab nebula pulsar was the first optical pulsar detected.

This led to the intriguing question: how is a pulsar formed? From all the evidence we have gathered we know that even a star cannot live for ever. Eventually it will exhaust all its nuclear fuel and then it will collapse. Fortunately for us it takes millions of years to burn itself out so there is no likelihood of our Sun suddenly flickering out leaving us in a cold and dark world—in spite of the fact that it is pouring out four million tons of its own mass each second. When a star eventually uses up all available lighter elements in its fusion processes, an implosion occurs. This sudden collapse creates such an increase in energy that the star throws off part of itself: an effect that we see across millions of millions of miles as a "nova"—the sudden brightening of a star for a short period before it settles back to normality. A star of about the Sun's mass can survive such an event but the tremendous gravitational attraction within more massive stars prevents a return to normal. Instead, the atoms are compressed to such an extent that their electrons are pushed into the nuclei. Such a star, although incredibly small, will still weigh the same amount. Its mass will not have changed; all that happens is that the space that normally

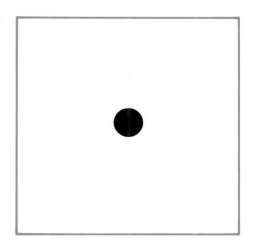

Above: the calculated size of a black hole of the same mass as the Earth. At a distance of two feet from a black hole of this size the force of gravity would be 100 million million times that felt at the surface of the Earth. An unwary astronaut approaching a black hole, it is calculated, would first be stretched like spaghetti before being trapped and sucked into the hole—where he would be crushed down into minute particles of matter.

Black Holes in Space

Below: Professor John Taylor, the British physicist and mathematician (seen here carrying out an experiment on the young Israeli psychic Uri Geller) has investigated the phenomenon of the elusive black holes. We cannot see their creation, he says, for even a star 10 times as massive as our Sun would fade from view in four millionths of a second at the moment of its transformation into a black hole.

exists within each atom—accounting for a great proportion of the atom's size—vanishes and the subatomic particles become packed tightly together. A matchbox full of such material would weigh a million tons. Stars that are not much heavier than the Sun eventually collapse to a size similar to the Earth's and are known as white dwarfs that is, old, very dense stars with nearly all their energy spent. They remain as heavy as they did in their larger state. It is the stars that are more than 1.4 times as big as our Sun that continue to collapse until they become neutron stars.

The pulsar found in the Crab nebula is particularly interesting because it seems to be a remnant of a very massive star that blew up in a "supernova" in A.D. 1054 and which was recorded by the Chinese. The nebula we now see, over 900 years later, is a great cloud of matter expanding in all directions. But the very core of the star remains, having collapsed, and is now flashing its existence to us from deep in space. Only one other visible pulsar has so far been found and that, too, appears to be the remnant of a supernova that occurred between 6000 and 10,000 years ago.

Confirmation that the theoretical neutron star, or white dwarf, really exists was a very exciting discovery and it has brought science face to face with another theoretical model which, if confirmed by observation, would be surely the most awe-inspiring object imaginable in the Universe: the black hole.

Although a neutron star is one that has collapsed to an astonishing density, even it reaches a point—when it is approximately 10 miles in diameter—when the nuclear forces within it prevent any further decrease in size. But what happens to much greater stars? When they reach the neutron star size during their implosion their gravitational forces are far stronger than any stabilizing internal energy and they rapidly grow smaller and smaller until they vanish. In place of the once massive star is a tiny particle that would still have the same mass as it possessed in its earlier state. But now its gravitational force is so strong that nothing can escape its influence, not even light! No matter how bright the star is we cannot see it. But if, in a spaceship, we went too close to the black hole its tremendous gravity would reach out and grab us and we, too, would vanish as we were sucked in and crushed to almost nothing.

Inside the black hole space and time would have very strange properties if we were to survive long enough to notice them and Einstein's Theory of Relativity gives us an opportunity to peer inside, theoretically, without being annihilated. But Einstein cannot be blamed for "creating" black holes, for though one lurks in his theory, Newton's law of gravity can also be said to allow such a phenomenon to exist. Indeed, the French scientist Pierre de Laplace observed as far back as 1789 that a massive condensed star would be invisible because light would be unable to escape from its surface. It seems that any massive star that cannot throw off most of its mass as it collapses will become a black hole, and since an estimated 90 percent of stars in the Universe are of the type that could become black holes, the chances are that outer space is littered by these invisible traps for the unwary traveler.

Not surprisingly, then, astronomers and scientists are putting a lot of effort into the search for black holes and, though they have not yet succeeded in identifying one with certainty, a number of excellent candidates are now the subject of intense study. Because they are invisible, their existence can be established only by the effect they have on their stellar neighbors. It is not possible for us to see the creation of a black hole, either, for as Professor John Taylor, British physicist and mathematician, explains in his book *Black Holes, The End of the Universe?*, a star 10 times as massive as our Sun would fade from view in four millionths of a second. Even a star a million times heavier than the Sun would vanish in a quarter of a second. This sudden disappearance would be impossible to observe.

One way of picturing events around a black hole is to go back to the rubber sheet analogy of space-time, in which the weight of objects causes depressions or curvature of space. Although it should not be taken literally, it is helpful to visualize the fantastically heavy particle that creates the black hole causing a deep narrow pit in the sheet. Travelers or stellar objects approaching the pit will feel the great gravitational pull and will notice time slowing down in the way already described for fast space travel, but if they have enough power they can escape from the black hole before it devours them. There is, however, an area around the hole called the *Schwarzschild radius*—named for Karl Schwarzschild, the German astronomer and physicist who first defined the barrier—and nothing crossing that point can escape being drawn into the center of the black hole. The radius also marks the end of visibility to anyone trying to look into a black hole.

If we assume there are in existence hundreds—perhaps even millions—of black holes, the majority of them will have been formed by stars which, at the outset, were no bigger than a thousand times the size of the Sun. In that event, as we travel through space, detectors on board our star ship will tell us of abnormal gravitational forces long before we cross the threshold of a black hole and we can avoid it. Even if we wanted to enter such a black hole its great force of gravity would crush us to death long before we reached the point of no return. But supposing a complete galaxy had, for some reason, imploded in the same way and was also a black hole (and it would be unwise to rule out any such possibility in our astonishing Universe). The situation would then be a very different one. They would be really dangerous black holes as far as space travelers are concerned. A spaceship would pass through the Schwarzschild radius (sometimes also called the "event horizon") of a galactic black hole without even feeling a bump and would be well inside before it noticed any difference in the pull of gravity. And yet it would not be able to escape. Just what would happen then we cannot be sure but a number of theories have been advanced. One factor that would influence the fate of the spaceship and its occupants is whether the black hole in question is rotating or at rest. It is possible that the spaceship may circle forever close to the Schwarzschild radius with time seeming to advance normally though if observers from outside were able to peer in the clocks on board would seem to be virtually at a standstill. The spaceship may be drawn to the center and crushed, of course. But a third alternative offers a

Below: a black hole and its distortion of space. It is, in fact, a difficult concept to illustrate, for the black hole would in reality be spherical and could be passed around in any direction. But at whichever direction it was approached, the space distortion (at the top of the diagram) would be met. If the approaching object—a spacecraft, say—could not counter the enormous gravitational force from the hole before it reached the point of no return (the broken circle that marks the Schwarzschild radius, or event horizon) then it would be sucked down to oblivion. From the event horizon to annihilation, not even light can escape the incredible force of gravity.

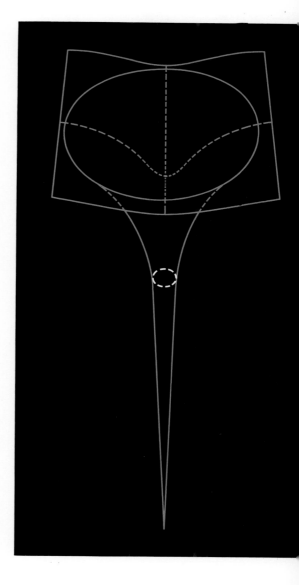

The Search for the Black Hole

Opposite: a model of a black hole in the x-ray source Cygnus X-1. Working from the laws of Newton and Einstein, physicists had predicted the existence of black holes, but it was not until the 1960s that the challenge to verify the theory was taken seriously. By the 1970s, scientists had decided that the most likely candidate had to be a star with a close companion that emitted x-rays. A black hole in a close binary system, they thought, might pull gas off its companion through its enormous gravitational attraction. As the gas fell into the hole it would heat up and emit x-rays. These were the facts that fitted a giant primary star in the constellation Cygnus and its dark companion Cygnus X-1. In the model opposite A shows gas being drawn off the giant primary star on the left by the gravitational attraction of the black hole. The hole is in orbit around the primary star and its movement causes some of the gas to miss the hole. The gas captured by the hole spins around in a flattened disk (blue). B shows the flattened disk of gas around the black hole. Pressures in the swirling gas cause a central bulge. C shows the vortex of swollen gas, caused by heat from the x-rays (red) given off near the black hole. D represents the core of the disk where the gravity is so strong that the bulge of hot gas is flattened. E: in the central 120 miles of the core, the pressure and heat cause the gas to give off the x-rays that can be observed on Earth. At around 50 miles from the center, the disk becomes violently turbulent and immensely hot. The disk terminates near the black hole where the gas, no longer able to move in orbit, is sucked into the hole.

more hopeful "way out." The extreme distortion of space within a black hole might create a "throat" that would be connected to another Universe. If that were so and the spaceship were able to travel through the throat without being squashed to nothing, it would be spewed out into a completely different Universe. But, of course, if the spaceship survived that experience those on board would not be able to communicate with us. They might, then, try to steer their ship back into the black hole that had regurgitated it in the hope that it would be ejected back into our Universe. But, according to the theory that conceived this idea, all that would happen would be that it would reappear in yet another Universe, and another, and another—but never the one from which it originally disappeared. But that *is* only a theory. The reality might be different and open up the possibility of using black holes as a means of rapid travel between various points within our own Universe. Some writers have even suggested that in the not too distant future scientists might even *make* a black hole sufficiently near our Solar System to be accessible without posing an immediate danger. In this way we could explore space and time in a way that is well beyond our wildest dreams at present.

Scientists might even try to make a small black hole within the confines of an Earth-based laboratory. If they did, according to Professor Taylor, the results would be disastrous. "A black hole weighing sixteen hundred tons, if let loose here, would immediately sink to the centre of the Earth and promptly devour it, and us included." And for good measure Professor Taylor declares elsewhere in his book that "even if we have only one black hole in the galaxy our ultimate fate is a gloomy one: we will be swallowed up by it in the end."

But before the black hole gets us—if it ever will—it may be possible to use its energy. Among the current theories on the subject, that of Stephen Hawking, a young Briton, has received much attention. He argues that the intense gravitational effects near the event horizon of black holes would split the constituents of atoms into particles and antiparticles. Under normal conditions they would annihilate each other, but the extraordinary effects of the black hole would swallow up one member of the pair allowing the other to escape. This could be detectable in the form of bursts of gamma rays. Hawking has also suggested that there might be many much smaller black holes which, due to the effects of this "leakage," are ready to explode with the force of millions of H-bombs. If that is so, says Hawking, man may in time find a way of harnessing this energy for his own use.

It may be only a few years before astronomers and scientists have indisputable proof of the existence of black holes. Then, from a study of these disturbing objects, a clearer picture will be formed. The period ahead, then, is one of intense study and consolidation of the knowledge we now have of the peculiarities of radio galaxies, the flashing pulsars, their non-radiating partners the BSOs, the mystifying quasars, and the astonishing, theoretical black holes. But by the time we attain a better understanding of these objects, new and more powerful instruments may well have introduced us to a range of even more puzzling cosmic phenomena.

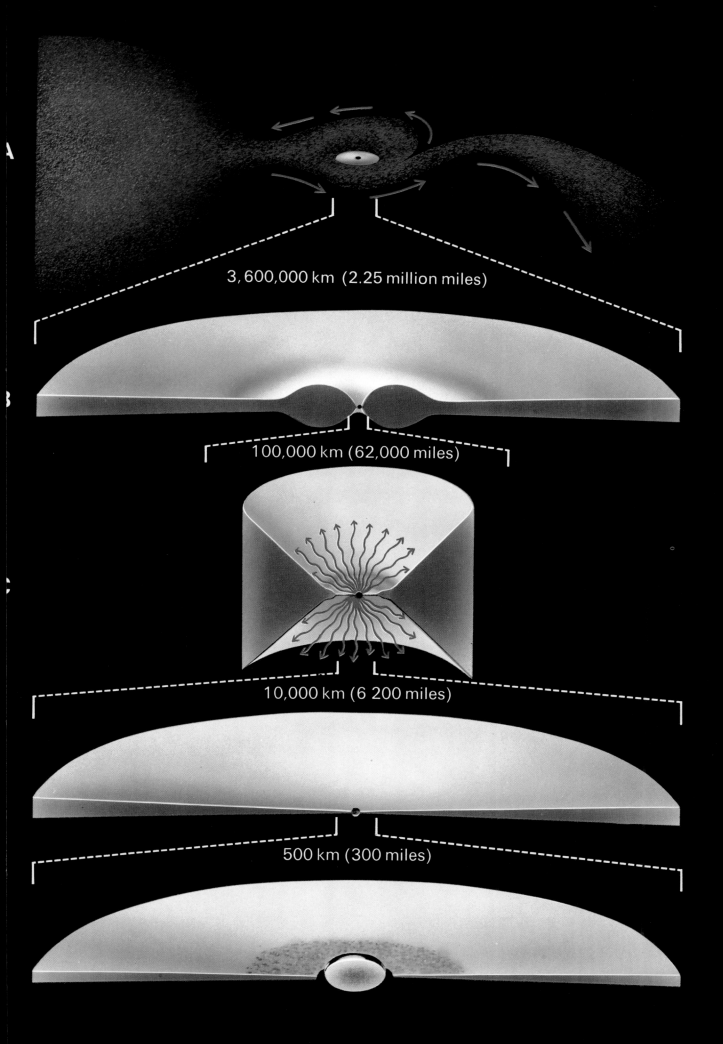

3,600,000 km (2.25 million miles)

100,000 km (62,000 miles)

10,000 km (6 200 miles)

500 km (300 miles)

Chapter 4
Worlds Without End?

Has the Universe always existed and will it continue for ever? Or did it have a beginning and will it also have an end? For centuries these questions were the province of philosophers . . . and it seemed that man would never be able to answer them with certainty. But, with the remarkable advances of 20th-century astronomy, some scientists now believe they know the truth about the origin of the Universe, and they also know its fate.

In the beginning, according to the Book of Genesis, God created the heaven and the Earth. In the beginning, according to 20th-century astronomers, there was a Big Bang. To show that the two views—religious and scientific—were totally compatible, Pope Pius XII adopted the big bang theory of the Universe as the official policy of the Roman Catholic Church in 1951.

Since the earliest days men had wondered how our stunningly complex world, and the vast, beautiful Universe in which it exists, were made. Simple, dramatic stories were invented to explain what happened and, all over the globe, they are remarkably similar. The creation myths of most races speak of the Universe as being either an egg or a formless ocean, out of which came a supreme being who brought order out of chaos and gave us light and life. These beliefs were embodied in the great religions that followed. And though the early Christians stifled much progressive thought about such matters, today's Church leaders regard scientific discoveries as further evidence of God's supremacy and influence in our lives.

It is fascinating to note that man, and later the Church, believed there *had* to be a beginning—a moment of creation. It is not surprising, then, that when leading astronomers in the 1940s stated their belief that the Universe *did* have a beginning, Roman Catholicism very soon approved the theory.

The big bang theory developed from the observations of galaxies in the early 1900s. Between them, three American

Opposite: an illustration taken from a 13th-century edition of the Old Testament showing God as the architect of the Universe. He carries out the planning with the aid of a pair of calipers—for even to the medieval mind there was an order and precision to all they could observe about the Universe. At this period, and until much later, the biblical account was not questioned and it was believed that the age of the Universe was comparatively short. James Ussher, the 17th-century Archbishop of Armagh, for example, gave the time and date of Creation as 10 a.m. on October 26, 4004 b.c., and this was accepted for many years.

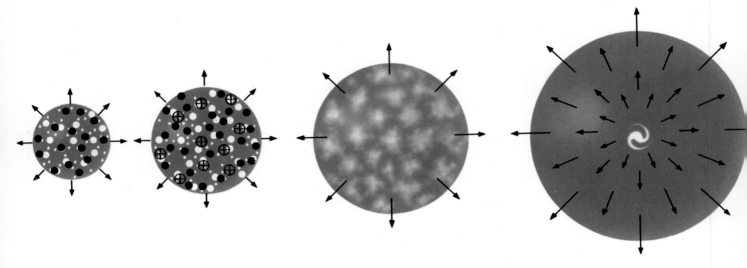

Above: a simplified view of the "big bang" theory of the creation of the Universe. Far left shows the "primeval atom," or cosmic egg, containing an incredible amount of matter packed tightly together. It contained also, an intense amount of radiation, which caused it to explode soon after the cosmic egg's formation. From this explosion the expansion of the Universe began. Second left shows the cooling of the embryonic Universe. Its particles combine to form the nuclei of atoms. Third from left, clouds of dust and gas form—the raw materials for galaxies.

Above right: another force, called cosmic repulsion, takes over and keeps galaxies moving apart.

Below: Georges Lemaître, the Belgian astronomer-priest who suggested the "big bang" theory of the Universe.

astronomers, Edwin Hubble, Vesto Slipher, and Milton Humason were able to show that not only were galaxies great "islands in space" made up of millions and sometimes billions of stars, but that they were moving away from us. Further examination showed that the galaxies were receding at increasing velocities. The farther away the galaxies were from us, the greater were the velocities: whereas the early investigators in this field had been surprised to find galaxies moving away from us at 400 miles a second, that very soon seemed to be a snail's pace compared with galaxy NGC 7619, which Humason studied in 1928. Its speed was 2360 miles a second. By 1936 even that seemed slow, for Humason then found galaxies with velocities of 25,000 miles a second, which is more than one-eighth the speed of light.

Why were these greater speeds being discovered? The answer was provided by Hubble, who was working with Humason. Using his colleague's velocity measurements and those of Slipher he showed that the speed of the galaxies increased proportionally with their distance from us. A galaxy that is twice as far away from us as another recedes at twice the speed. A galaxy four times farther from us (in any direction) than another will be speeding away at four times the velocity. This is known as "Hubble's law" and subsequent study confirmed these findings. The speeds involved were so enormous that some astronomers questioned the validity of the evidence. Was the shift in the spectral lines toward the red end of the spectrum (the Doppler-Fizeau effect, which enables scientists to determine the velocity and direction of distant objects) really caused by a rapid movement away from us, or could some other influence be at work? No satisfactory alternative could be found. The astronomers were forced to accept the evidence of their instruments and to try to fit the observations into a model of the Universe that also incorporated other astronomical knowledge. The big question was: why is everything fleeing from *us*?

The man who supplied the answer was Georges Lemaître, a Belgian astronomer, who suggested in 1927 that all the matter in the Universe was once condensed into one huge mass that became unstable and exploded. He called the original mass of

matter—which may have been only a few light-years in diameter
—the "cosmic egg." Lemaître's ideas come over clearer if we
imagine, instead of an egg, a balloon on which billions of dots
have been painted. The moment of explosion would correspond
to the moment we start blowing air into the balloon. Supposing
we were small enough to stand on one of the dots and, as the
balloon expands, measured the distances between us and the
10 nearest dots in all directions. Later, after the balloon had
expanded much more, we again measured the distances between
us and the same 10 dots. We would find that they were now all
much farther from us.

That is what is happening in our expanding Universe. Instead
of dots on a balloon we have galaxies in space, and no matter
which galaxy we view the heavens from we will always find that
the other galaxies are moving away. Observations in the 1920s

The Big Bang Theory

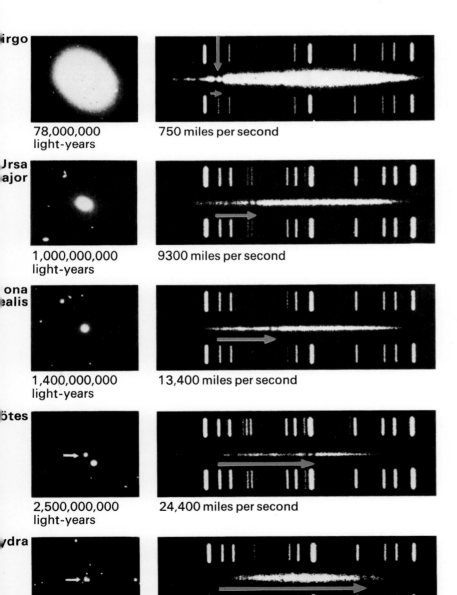

Virgo
78,000,000 light-years
750 miles per second

Ursa Major
1,000,000,000 light-years
9300 miles per second

Corona Borealis
1,400,000,000 light-years
13,400 miles per second

Bootes
2,500,000,000 light-years
24,400 miles per second

Hydra
3,960,000,000 light-years
38,000 miles per second

Left: examination of five star clusters from galaxies at different distances away from Earth. The figures seem to prove the discoveries of Edwin Hubble and Milton Humason, the American astronomers, that not only is every single galaxy (except some of those in our Local Group) moving farther and farther out into space, but that this outward speed—velocity of recession—increases the farther away a galaxy lies. Starting with a star cluster from the galaxy in Virgo (in top row, left) we see that it (and hence the whole galaxy) is 78,000,000 light-years away. Examination of its light in the spectroscope (top row, right) shows that it has a redshift when measured from a specific baseline—in this case the H and K lines of the calcium atom. This can be interpreted mathematically in terms of the Doppler effect and a recession speed of 750 miles per second given. The second star cluster, from Ursa Major is 1000 million light-years away. The H and K lines are redshifted even farther to the right, or red end of the spectrum, and the reading is worked out at 9300 miles per second. Thus Hubble's Law, that "the apparent speed of a receding galaxy is proportional to its distance from Earth" is shown to be correct.

Above: the balloon theory, often used in answer to the question: Why is it that all other galaxies are moving away from us? As a balloon expands, marks on its surface move farther apart. Marks that are already far apart, such as A and C in the illustration show a bigger increase in distance than marks that are close—A and B. Clusters of galaxies move apart in a similar way.

gave us speeds and distances of the galaxies with which science could attempt to answer for the first time the question: When was the Universe created? Knowing the rate of expansion enabled astronomers to work backward—like running a film back through a projector or letting the air out of the balloon. They calculated that all the matter in the Universe, in the form of a cosmic egg, blew up 2000,000,000 years ago.

Something was wrong with the figures, however. Geologists were certain from their study of the Earth, and in particular from the breakdown of radioactive uranium into lead, that our planet was at least 4000,000,000 years old. It takes 4500,000,000 years for half of any quantity of uranium to change into lead, and the dating of our planet's origin was based on the ratio of uranium and lead that presently exists. But the Earth could not be twice as old as the Universe, so either some aspect of the big bang theory was at fault or the geologists were wrong. This problem was not resolved until the early 1950s when Walter Baade, the German-born astronomer, revised the Cepheid yardstick and, overnight, the Universe was known to be at least twice as big as originally thought. This discovery meant that calculations for Hubble's law were also wrong, and when revised the moment of the big bang was found to be 13,000,000,000 years ago—a figure that is now widely accepted by astronomers. The Earth, then, did not form until the Universe was at least 8000,000,000 years old.

Although it fits the facts, the big bang theory leaves as many questions unanswered as the opening lines of Genesis. Who or what, to put it crudely, "laid" the cosmic egg? How did the matter come to exist in the first place? What determined its mass? How long had it existed before the fantastic explosion? And what was it that caused this huge chunk of matter to become unstable and blow itself into galaxies, stars, planets, moons, and men? These are questions that most cosmologists ignore because they cannot attempt to provide answers at present. But clearly the big bang was not *the* beginning, just *a* beginning, or a stage in the evolution of the Universe.

Nor is the big bang theory the only model that fits the events in our expanding Universe. There are many, but the most serious opponent—at least for some years— was the steady state theory first put forward in 1948 by Hermann Bondi and Thomas Gold, both British astronomers, and later extended by their well-known colleague Fred Hoyle.

They accepted that the galaxies were receding and that the Universe was expanding, but they argued that it did not have a beginning in the form of a big bang. Instead, they suggested that we lived in a "continuous-creation" Universe where matter was being formed all the time to take the place of those galaxies that, as they sped away, reached the speed of light and disappeared from our Universe. So, at whatever time you look at the Universe, it will seem the same and have the same density. Such a Universe could have existed for all time and could continue for ever. The chief objection from fellow astronomers was that the theory contravened the law of conservation of mass-energy, in other words it seemed to require matter to be made out of nothing. The answer was that the energy for the creation of the new matter

The Expanding Universe

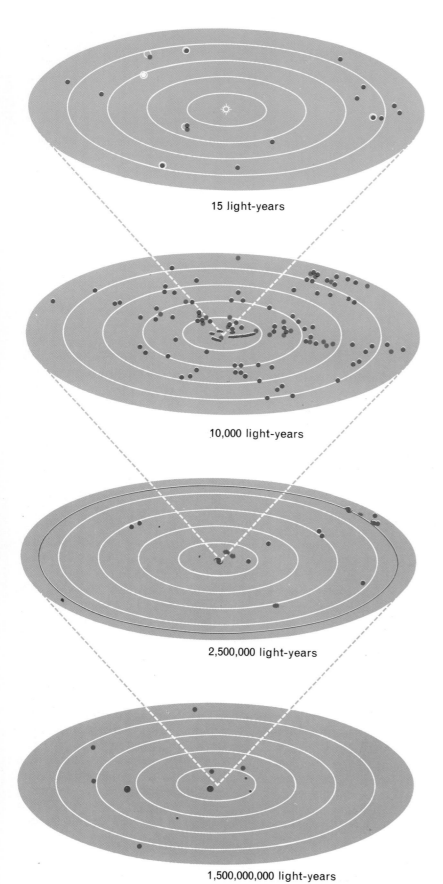

15 light-years

10,000 light-years

2,500,000 light-years

1,500,000,000 light-years

Left: this series of diagrams gives some idea of the vast distances involved when discussing the Universe. Top shows the Sun and its nearest neighbors within a distance of 15 light-years. Those stars ringed in a white circle are visible to the naked eye. Colors of stars represent their surface temperature. Purple: 3000°C; red: 4500°C; green: 5500°C; white: 10,000°C.

Second left: the Milky Way extending out to 10,000 light years, with the Sun and its nearest neighbors at the center. Green dots represent sources of bright hydrogen emission. Red blotches are obscuring clouds, and purple dots represent young stars and open clusters.

Third left: the Local Group of galaxies and nebulae. The purple spiral at center is our Milky Way with the Magellanic Clouds nearby. The red circle represents the limit of observation with the naked eye from Earth. At the limit Andromeda—M31— can just be seen.

Bottom left: many other galaxies lie beyond our Local Group, even when we extend our "horizon" to only as far as the Corona Borealis at 15,000 million light-years. Red dots represent clusters of galaxies of over 400 members. Green dots represent clusters of between 100–300, and purple dots clusters of less than 100—the Local Group comes into the purple dot category. All the clusters here are receding from us at hundreds of miles per second.

The Steady State Theory

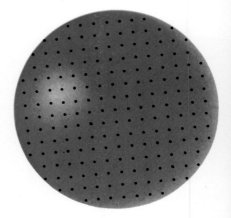

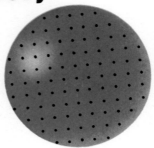

Above: an illustration of the "steady state" theory of the Universe, in which the galaxies are always equally spaced because new matter is constantly being added. This was the most serious of the alternative theories to the "big bang" explanation of the Abbé Lemaître. According to Herman Bondi, Thomas Gold, and Fred Hoyle, the British astronomers, the Universe did not have to begin with an explosion. They suggest that although the galaxies are moving farther away from each other, new matter is always being created. From this new matter, which takes the form of hydrogen atoms, new galaxies are created. In this theory, the total amount of material in any large region of the Universe never varies.

Below: Fred Hoyle, who reconciled Bondi and Gold's theory of the steady state Universe with the idea of a local "big bang" for a smaller part of the Universe

could be derived from the expansion of the Universe. Since it would require only one atom of hydrogen to be formed each year in every 200,000,000 gallons of space, it was hardly surprising that it had not been detected, and besides, how could we be sure that the law of conservation of mass-energy (which was seen to apply on Earth) applied everywhere in the Universe?

If we were to run back the "film" of such an expanding Universe—as we did with the big bang model—the matter in the Universe would not condense into a great cosmic egg. It would gradually break down into hydrogen atoms that then slowly vanish, throughout the Universe, as the galaxies come back over the "horizon."

So, by the end of the 1950s, astronomers had two very credible theories to explain why the Universe looks and behaves as it does. Which was right? Despite its attractiveness, the steady state theory has been abandoned by many astronomers in favor of the big bang, and even Fred Hoyle became convinced in 1965 that the big bang was the right explanation, on a "local Universe" basis: that is, that we are caught up in the aftermath of a small explosion in a far larger steady state Universe.

One of the most enthusiastic supporters of Lemaître's theory was George Gamow, the Russian-American astrophysicist who, incidentally, gave it the name "big bang." Gamow conjectured that the early Universe consisted largely of high-intensity radiation rather than matter. As it expanded, the radiation's intensity dropped rapidly with the result that we now live in a matter-dominated Universe. As a remnant of the big bang, Gamow argued, a faint background of radiation should still be detectable.

Among other theorists who predicted much the same thing was Robert H. Dicke of Princeton University. He said during the 1960s that if everything began with a huge explosion there would have been a great flood of energetic radiation, such as X-rays and gamma rays. If we could delve into the farthest regions of space, looking back over thousands of millions of years, we should still be able to detect it, though the distance would cause this radiation to be red-shifted into the microwave region. In that case, such a faint background radiation should be found in whatever direction a radio telescope looked. And that is exactly what has been found.

The discovery, like so many in science, came about accidentally. In 1965 Arno Penzias and Robert Wilson, two scientists at

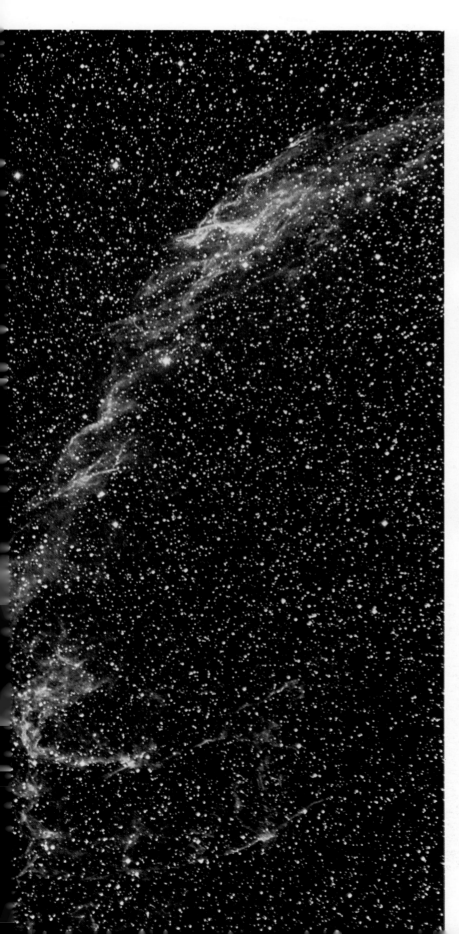

Top, Herman Bondi, and above, Thomas Gold produced their theory of the steady state of the Universe in 1948. As matter recedes beyond the limit of observation, they said, newly created matter takes its place.

Left: the Veil nebula in Cygnus. The bright area is caused by shock waves from a supernova explosion that ionize particles of gas and give off radiation in the form of light. It is probably in the dust and gas of nebulae that most new stars are born.

The Echo of the Big Bang

Below: Arno Penzias (right) and Robert Wilson (left), the American scientists who in 1965, while working at the Bell Laboratory in New Jersey, discovered and identified a faint background noise in the electromagnetic range coming from all directions of outer space. This radiation was named "cosmic background radiation." It has provided astrophysicists with another valuable piece to add to the jigsaw puzzle of the origins of the Universe. Cosmologists (scientists whose task is to make a convincingly coherent picture of the Universe and uncover its history) now believe that this was the intense radiation present at the time the "cosmic egg" exploded, and that the faint noise now being received is the "echo" of the considerably cooled radiation particles in the farthest reaches of space.

Bell Telephone Laboratories in the United States, were making radio-astronomical measurements using instruments designed to pick up signals reflected from satellites. After accounting for all sources of radiation they found there was a residual amount that they could not explain. Furthermore, it had a strange characteristic—it was detectable from every direction. This radiation is interpreted by most astronomers as an "echo" from the big bang. Its temperature is three degrees above absolute zero (3°K) and it is the strongest evidence in favor of an explosive start to the Universe.

The discovery of quasars is one which, in time, may put yet another nail in the coffin of the steady state theory. During the 1960s these huge and very luminous objects (up to 100 times as bright as our own galaxy) were found to be receding from us at great speed toward the edge of the observable Universe—at least, that is the present interpretation of the observations. Some astronomers argue that they may not be so far or traveling so fast and that the red shifts recorded for the quasars are caused by other factors, such as intense gravitational fields. If that turns out to be the case, then they might be much closer to us and more evenly distributed. But if further investigation confirms that the quasars are what they first seemed to be then it would show, contrary to the arguments of the steady-statists, that the Universe is not the same at all times. If the bright and massive quasars are found to exist only very far away, then we are seeing them as they were thousands of millions of years ago. That indicates that in the early stages of the Universe there existed objects that are no longer part of our local Universe.

Like clothes, scientific theories come in and out of fashion, and some scientists suggest that the big bang theory *is* just a fashion and that it will eventually be discarded by a new design or an existing one that suddenly appeals once more. Jayant Narlikar, professor of astrophysics in Bombay, India, and a collaborator with Hoyle on a theory of the Universe, has this to say about the steady state theory in his book *The Structure of the Universe:* "This theory makes certain very definite predictions, and must be abandoned if these are not borne out by observations. The theory has been attacked, and even claimed to be disproved, from time to time, with the help of data which has either been subsequently withdrawn or proved inconclusive. . . . So far the only genuine difficulty faced by the theory is that arising from the microwave background. If no alternative sources (to big bang) of this background can be found, the steady-state theory must be abandoned. Yet the popular belief, shared even by most astronomers, is that the theory has already been disproved once for all."

So, bearing in mind that the big bang theory may not, ultimately, prove to be the right one, let us now apply it to the Universe and see how the great Creation occurred. We will have to ignore questions about the origin of the cosmic egg at this point, apart from making the observation that it seems to be as much a violation of the conservation of mass-energy law as that of the steady-state theory. We must also assume that only known laws of nature came into force.

According to a popular picture of the moment of creation,

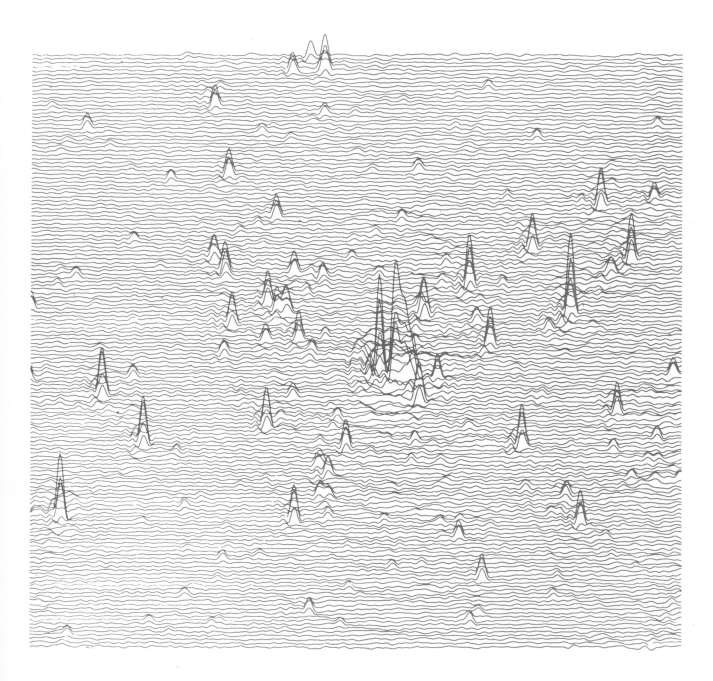

the big-bang produced a huge intensely hot fireball of radiation. It cooled to a million million degrees within a thousandth of a second and then the various cosmic forces came into play as it continued to expand rapidly. In those early stages the radiation was so intense that matter was kept in a frenzied state. But after hundreds of thousands of years the fireball cooled sufficiently for atoms to form and by the time it was at about the temperature of our Sun—5000°K—matter was able to "disengage" itself from the radiation, which continued to cool to its present 3°K. This happened when the Universe was about 300,000 years old and only about one two-thousandths of its present size, and still densely packed. Gravity then began gathering the matter together into galaxies in which individual stars formed—the first appearing perhaps 1000 million years after the big bang.

Above: radio sources chart of a small patch of sky above Cambridge, England, made with the one-mile telescope of the Mullard Radio Astronomy Observatory at Cambridge University. The sets of peaks on the chart represent a radio galaxy or a quasar. There are, however, a large number of other, fainter radio sources caused by cosmic background radiation being transmitted in the radio region of the spectrum. Although faint on the chart they are actually powerful sources at great distances—distances so great in fact that by the time we receive them on Earth they have long since disappeared.

Right: a 17th-century engraving illustrating
the biblical concept of the cosmos as a
chaos in which hot, cold, moist, and dry
elements are locked in struggle.
Below: the arguments for both big bang and
steady state theories of the Universe are still
inconclusive. Here is yet another theory—
that of the oscillating Universe. An initial
cosmic egg expands to its maximum size
during the course of some 45,000 million
years. It then contracts and reforms into a
cosmic egg once again. The cycle then
repeats itself.

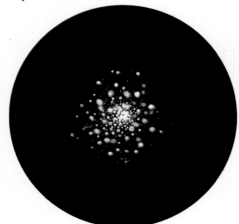

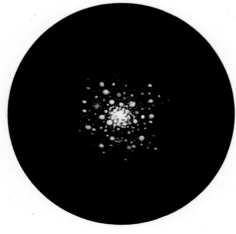

Inside the stars nuclear reactions were taking place, converting
hydrogen into helium and releasing energy. The more massive
stars burned rapidly and then exploded, scattering elements like
carbon and oxygen, as well as uranium, into space. Gravity
took hold of this enriched debris and shaped it again into new
stars and planets. Our Sun and Earth were formed in this
way. (The atoms from which we, our planet, and the entire
Solar System are made were once in another star, much more
massive than our Sun, which probably blew apart in a supernova.) Eventually, when favorable conditions on Earth came
about, life appeared in many forms.

That, in very simple language, is how scientists believe the
Universe began, even though there are many events contained
within that theory that so far defy adequate explanation. One
of the most perplexing is the way galaxies are formed. Why was
it that the big bang did not throw out matter *evenly* in all directions so that the stars are the same distance apart? The fact that
stars travel through space in huge groups, and that even the
galaxies travel in clusters, indicates that the initial explosion was
somehow "lumpy."

But if we accept the big bang theory, spanning 13,000,000,000
years from the explosive birth to our present stage of evolution,
what does the theory tell us about the future? Will the Universe
go on expanding for ever, with everything moving steadily
farther apart from its stellar neighbors? Or will the expansion
slow down and come to a halt or even go into reverse, with all
the matter eventually returning to form a great, dense cosmic egg
once more?

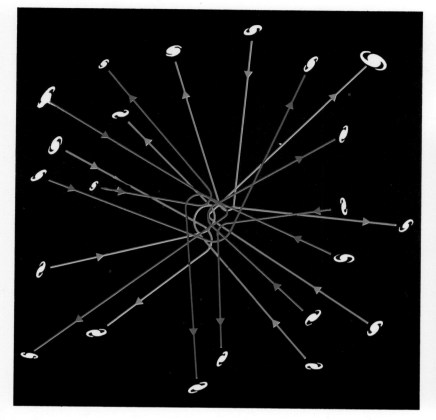

Theory of the Oscillating Universe

The theory tells us that the deciding factor is the amount of matter in the Universe. If it is below a certain level—about one atom per cubic meter—then it will continue expanding forever. But if there is more than the critical amount then the forces of gravity will eventually prevent further expansion, causing a recollapse. Finding out just how much material there is, however, is far from simple and it is a subject of heated debate among scientists. Some scientists, for example, believe the bulk of a galaxy's mass may be "hidden" in a halo too faint to be detected on ordinary photographs. Indirect evidence for an "open" Universe (low density) is to be found in the recent discovery of deuterium (heavy hydrogen) in interstellar space. If, as some theorists suggest, it was formed in the big bang then a low density Universe is necessary to account for its survival. That means the Universe will expand for ever. Further evidence for an ever-expanding Universe may have been provided by Dr Arthur F. Davidsen, an American astronomer at Johns Hopkins University, who announced in April 1977 that he had been able to photograph a quasar at the very edge of the observable Universe with a telescope fixed to a rocket. The picture was so sharp that he concluded there was insufficient matter between the galaxies for the Universe ever to be pulled into contraction by the force of gravity.

In addition to the stars and galaxies there is an abundance of other material in the Universe in the form of gas and dust. We, in the spiral arm of the Milky Way galaxy, are in an area of debris that obstructs our view of the Universe in some directions. The Horsehead nebula and the Northern and Southern Coal-

Above left: diagram of an expanding and contracting Universe developed by Professor Hannes Alfvén (above) the Swedish physicist. Alfvén, working on a theory originally put forward by Oskar Klein, pointed out a fallacy in the big bang theory of the Universe where it relied on the fact that all the galaxies, which are receding from each other at increasing speeds, were therefore once all together. By mathematically "winding back" the galaxies he showed that they would not, in fact, have all been in the same place at the same time. Klein and Alfvén's theory was that an immense cloud of gas began to contract under its own gravitational attraction. During this contraction a succession of nebulae condense and move relative to each other in hyperbolic orbits. They eventually fall back into the center and diffuse again. What observers on Earth are witnessing is the expanding phase.

Birthplace of the Stars

sacks are very impressive regions of dense dust that prevent us from seeing visible light beyond. Estimates of the amount of dust scattered throughout the Universe vary greatly, though the suggestion that the dust in our own galaxy is sufficient to make up between 2–3000 million stars is regarded as "the lowest estimate." Some other galaxies, such as the Large Magellanic Cloud, might have three times the concentration of interstellar matter.

This, of course, is matter that can be seen or detected. But if black holes exist, then they might have swallowed up matter that we can no longer see but which is still part of the Universe. Indeed, there might well be many black galaxies. If so, their existence would have an important bearing on the question of the expansion or collapse of our Universe. Professor John Taylor remarks in *Black Holes: The End of the Universe?* that the elliptic galaxy M87 "may contain as much as 98 percent of its matter in a form we cannot see. It is natural to suggest that it occurs in the form of black holes, obviously invisible to us except by the extra energy they impart to the neighbouring visible stars in their galaxy." Similarly, some clusters of galaxies have been found, says the professor, that have a high proportion of dark matter, "usually between ten and a hundred times that seen." A case in point, he says, is the Virgo cluster with about 50 times more invisible than visible matter.

Though some astronomers believe that it is the fate of the Universe to expand for ever, much more evidence is needed about the quantity and distribution of matter before the issue can be resolved, and the black hole theory may prove to be crucial to our understanding of the total mechanics of a big bang Universe. This will be particularly so if—as some scientists now believe—black holes are found at the center of galaxies. There may even be one at the center of our own galaxy. That, at least, seems to be one of the implications of remarkable experiments being carried out at the University of Maryland by Joseph Weber, the American astronomer. It had been suggested that a massive star collapsing into a black hole would produce detectable quantities of gravitational radiation. To test this theory Weber set up five-foot long cylinders suspended by wires in a vacuum. Quartz crystals were bonded to the surface of these aluminum cylinders to give them incredible sensitivity. If gravity waves existed the apparatus should be able to detect them. Indeed, the cylinders were so sensitive that they picked up many other things too, such as vibrations from marching students. In order not to be misled by local vibrations another detector was set up 620 miles away near Chicago. If oscillations were detected simultaneously by both detectors this would be strong evidence, said Weber, that they were caused by an extraterrestrial source of radiation.

Weber's results were impressive. In 1969 he announced several hundred observations over several months that could not be dismissed as chance fluctuations. The oscillations were strongest when the detectors were pointed toward the center of our galaxy. There was also a detectable pattern: a short pulse lasting less than half a second that occurred once every four days, at a frequency of around 1600 cycles per second. What was causing it? Professor Taylor offers this explanation: "Brief

Opposite: the Orion nebula, some 1500 light-years away is a region where new stars have just been formed. Close study of nebulae can reveal much information to cosmologists. Much of the gaseous medium that makes up a nebula consists not only of the raw material out of which new stars are formed, but also matter that was once in the interior of stars and has been returned to interstellar space. They can be thought of in some ways as gigantic recycling processes.

The Search for Gravitational Radiation

outbursts of radiation of such frequency contain a considerable amount of energy and if they are produced from the centre of the galaxy would correspond to it emitting up to two hundred solar masses as gravitational radiation each year. That is an enormous amount of radiation, and we are left with the disturbing problem of explaining where it came from. It certainly would not be satisfactorily explained by the final collapse of a star which had imploded as a supernova, for that happens only once every hundred years or so in our galaxy; Weber saw it happening once every four days." One explanation could be that the radiation "is coming from stars near the galactic centre as they fall into a large highly spinning black hole which forms the central part of our galaxy. The mass of this terrible nucleus could be as high as 100 million times that of the Sun, and it is gobbling up the stars on its outer rim at between one and 30 solar masses each year."

But scientists are still treating Weber's research with great caution, particularly as others have not been able to duplicate his results and because some suspect that other causes, such as seismic activity within the Earth, may be responsible for some if not all of the oscillations. Others have accused him of "subconscious bias" and so, to rule out that possibility, Weber arranged for his assistants to carry out a series of experiments and declined even to touch the instruments involved. In a letter to the scientific journal *Nature* in March 1977, Weber answered his critics by describing tests conducted without his direct involvement. The assistants did everything, including selecting certain random data, and positive results were still produced. But before science accepts these findings, other researchers will have to confirm them with their own tests, or similar equipment operating in space or on the Moon, will have to pick up the oscillations simultaneously with instruments on Earth.

Even if Weber fails to find proof of a black hole at our galactic center, the nuclei of galaxies will be the subject of intense research in the next few years—particularly those nuclei that seem to have peculiar properties. Many are certainly the scene of intense and violent activity. An arm-like structure consisting mainly of atomic hydrogen extends for about nine light-years from the center of our own galaxy and is moving toward us at 30 miles a second.

The galaxy known as NGC1086 (the letters stand for New Galactic Catalog) is a weak radio galaxy—that is, its microwave emission is only about 100 times that of a normal galaxy. But what puzzles astronomers is that the microwaves come from a tiny source at the very heart of the immense galaxy. Could it be that we are seeing the beginning of a great explosion? If so, the events in another galaxy, M87, may be a later stage of a similar catastrophe. There is an intense source of microwave emission at the heart of M87—the most massive galaxy that we have detected—but also a weaker halo of radiation spreading out from the center. A luminous jet can also be seen optically shooting out from the center of M87. A later stage still may be when the microwave radiation passes through the galaxy and emerges on either side. In the case of another galaxy (NGC5128), which appears visually to be split in half by a band of dust, there are two intense radio sources on either side of the central band, but also

Below: a supersensitive aerial in the form of an aluminum cylinder used by Professor John Weber, an American astronomer, to detect gravitational radiation. When Weber trained the cylinder in the direction of the center of the Milky Way he received impressive readings of outbursts of radiation activity. Are the readings, as one scientist suggests, the dying cries of stars as they are gobbled up by a large, highly spinning black hole at the center of our galaxy?

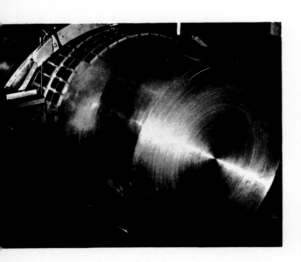

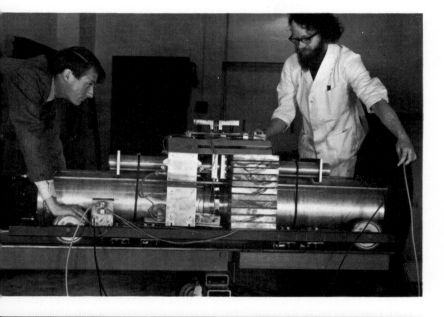

Above left: gravitational radiation detection equipment set up at Glasgow University in Scotland. Intense interest in Weber's experiments have led other scientists to try to duplicate his results. However, many scientists are not satisfied that present-day detection equipment is sensitive enough to verify the findings.

Left: American astronaut Edwin Aldrin setting up equipment on the lunar surface to relay information about moonquake activity back to scientists on Earth. Because the Earth is too "noisy," Weber's gravitational radiation results are difficult to interpret accurately and the Moon has been suggested as a quieter base (little natural, and no man-made "noise") to conduct gravity radiation experiments.

Exploding Galaxies

Below: two photographs of the galaxy M82. The upper photograph shows the distribution of stars and interstellar dust and gas. The lower photograph, taken with a special filter, shows a gigantic plume of interstellar material—hydrogen— exploding from the galactic nucleus. Fainter plumes can be seen that suggest earlier explosions.

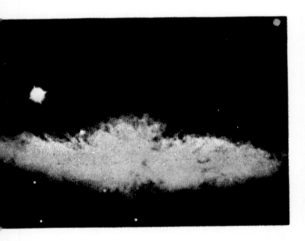

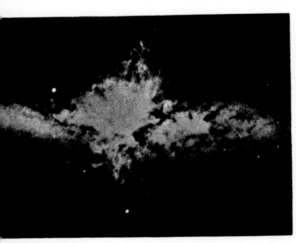

two much larger areas of weaker microwave emission on either side of the visible galaxy. Three-quarters of all known radio galaxies present a similar picture to us.

More information about exploding galaxies came in the early 1960s when Clarence Lynds, an American astronomer, was searching for an optical counterpart for a radio source known as 3C-231. He pinpointed it as a small galaxy, M82, which has unusual amounts of galactic dust that prevents individual stars from being seen—even though it is comparatively close to us. When viewed through a 200-inch telescope by Allan Sandage, the American astronomer, however, it proved to be a very interesting object. Having reasoned that something in the center of the galaxy may be throwing out matter, and that this matter would be mainly hydrogen, Sandage decided to photograph M82 using three-hour exposures through a special red filter that would let through light associated with hot hydrogen. His photograph confirmed that the galaxy was in the grip of a colossal explosion. Jets of hydrogen up to 1000 light-years long were to be seen stretching from the nucleus. They contained enough matter to form 5,000,000 stars or more. The immense explosion must have started 1,500,000 years ago, and yet it seems to be in an early stage of development since it has not yet developed a double source of microwave emission on either side of its center.

One particular type of galaxy, named for Carl K. Seyfert, the American astronomer and astrophysicist who discovered them in 1943, is particularly compact at the center and is unusually hot and active. One out of every 100 galaxies may be of this type. What has excited interest is that their nuclei are very much like stars, showing the same broad emission lines. They also emit strong infrared radiation. Similar to the Seyfert galaxies are N-galaxies—those that emit strong radio waves and have large red-shifts. Supposing these Seyfert and N-type galaxies were much farther away from us; what would they look like? Not like galaxies but like enormous bright stars. Could it be, then, that the so-called quasi-stellar sources—quasars—are not strange new objects at all, but simply distant Seyfert or N-galaxies? They are so far away that all that we can see at present are their bright starlike centers. If that is so, the quasars would not after all be evidence against the steady-state theory of the Universe.

The puzzle that remains is why so much violent activity is going on at the heart of so many galaxies. Is it a stage that many galaxies pass through? Will ours explode, or has it already done so? Evidence taken from the Earth's crust and the existence of neutral clouds of hydrogen in our galaxy indicate to some scientists that the center of our galaxy did explode many millions of years ago.

If we were to understand more about the "small bangs" at the center of galaxies, they might shed light on the "big bang" of the Universe, if similar processes are at work. Although the big bang and steady state theories are the two models that have most occupied the minds of astronomers, physicists, and theorists, there are many alternatives to, and modifications of, these ideas. It could be that, within a decade or two, our picture of the Universe will be very different to the views we now hold.

One of the alternatives to the big bang and steady-state

concepts is that antimatter plays an important role in the creation of a Universe and that the Universe we now see is made up of equal amounts of galaxies and antigalaxies. Another view is that the Universe pulsates—expanding and contracting—with the matter and antimatter coming together then exploding again into two separate Universes. Yet another possibility is that while the Universe of matter is expanding, its twin anti-matter is contracting, and vice versa. A pulsating Universe is possible, of course, without recourse to antimatter. Indeed, the big-bang theory may not be describing a beginning for the Universe but the beginning of a pulsation, one of perhaps hundreds, thousands, or billions of which may have occurred in the past. With each contraction all the matter condenses, to be reborn again in a new Universe. Yet another possibility involves

Below: map made by Dutch and Australian radio astronomers showing the distribution and density of neutral hydrogen in the Milky Way. Neutral hydrogen in interstellar gas is undetectable by optical astronomy, but since its discovery through radio astronomy, there has been much speculation that our galaxy, too, has passed through a period of violent activity—possibly even a gigantic explosion.

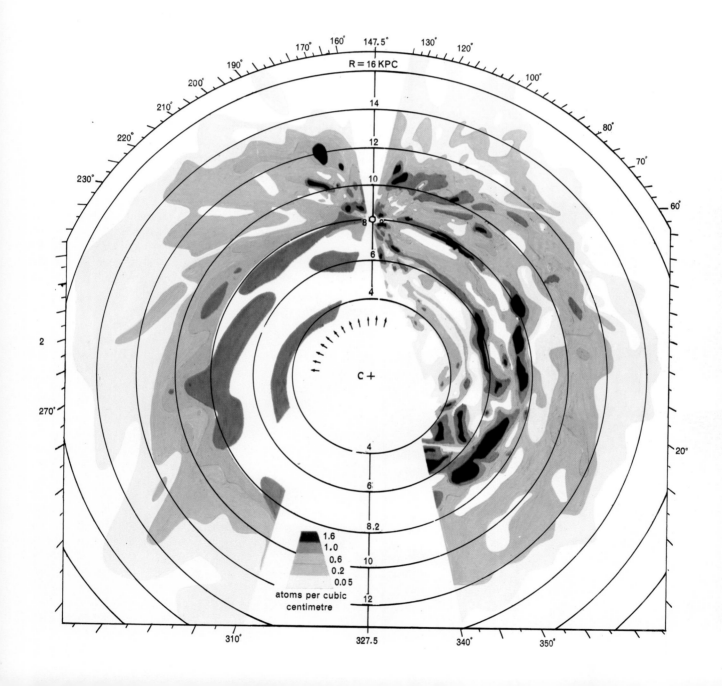

Telescopes in Space

Below: the skylab space station in action. It carries eight telescopes and various sensors, which are invaluable aids to astronomers. The thin envelope of water and gases that is the Earth's atmosphere cuts off most of the rays coming to us from outer space and limits the vision of even the most powerful Earth-based telescope. The helicopter-vanelike apparatus is the telescope mount, with a four panel solar array to generate the skylab's electrical power. Behind the skylab's octagonal "nose" is the orbital workshop—a forward compartment containing food and equipment storage, and a crew compartment. A solar array wing at the front left was ripped off during launching.

the ubiquitous black holes. Just as black holes are thought by some astronomers to be supplying the motive power at the center of galaxies, other theorists suggest that there might be a black hole at the center of the Universe. Or, to be more precise, there might be a black hole in another Universe into which matter is falling and being spewed out into ours. The cosmic egg, then, may not have been a lump of dormant, condensed matter that was suddenly made unstable, but a "white hole"—the end of a funnel through which matter from another Universe could pour and take a new form. What would happen to our Universe if it did collapse after a period of expansion? Would all the matter collect in a cosmic egg or would its size cause it to continue collapsing, like the massive stars, until it became a black hole, ejecting all the matter into yet another Universe? The possibilities are almost endless and it will be a long time before the theorists and astronomers can explore the various models, and confirm or dismiss them with observations.

The American Space Shuttle program will play an important role in the 1980s, for there are plans to take into space various pieces of astronomical equipment with which the Universe can be studied well away from the limitations imposed by the Earth and its atmosphere. While we await these developments and ponder the implications of what we now know about the Universe, it is worth reflecting that although it seems at times to be a very violent place there are cosmic forces that keep every-

thing—from subatomic particles to the massive galaxies—in order. And the intriguing case of the "coincidences of large numbers" has given many scientists food for thought. When the electrostatic force between an electron and a proton is compared with their gravitational force, the former is found to be about 10^{40} times larger than the latter. If we compare the radius of the observable Universe with the radius of the electron (though neither are rigid spheres) we find the first is 10^{40} times greater than the other. Finally, the same number, 10^{40} is roughly equal to the square-root of the total number of particles in the observable Universe. Coincidence? Or are they evidence of an overall plan whose design we are only just beginning to understand?

Throughout history we have had to continually expand our picture of the Universe, which now reaches out almost to the limit of the observable Universe—which seems to be about 3.4 thousand million light-years in all directions. Is that all there is? Could it be that once we have discovered beyond doubt how our Universe was created we will discover that much more lies beyond: that we are just one of many Universes that explode and expand, contract and disappear, or even collide and annihilate each other like the stars and galaxies we are observing in our own region of space.

Perhaps the galaxies are like subatomic particles and our Universe is a single atom in a far greater scheme where there are worlds without end—or beginnings.

Above: artist's impression of a delta-winged Orbiter—the re-usable payload of the Space Shuttle. It will orbit the Earth and carry out various tasks such as carry a giant telescope above the atmosphere. Its mission completed, Orbiter will re-enter the atmosphere and land as a jet plane. Below: an Orbiter mounted on a Boeing 747 prior to its atmospheric test trials.

Chapter 5
The Enigmatic Solar System

What do we know about our closest neighbors in space, the planets that also orbit around our sun? Over the centuries, theories and speculation have abounded—tales of a system of canals on Mars, visions of Venus as a possible cradle of life under its thick cloud cover. The reports we receive from space probes suggest that neither Mars nor Venus are hospitable to life. But what are they like? And what about our farther neighbors? Will it ever be possible for a spacecraft to glide past the beautiful rings of Saturn to see the planet itself? What are the answers that the astronomers are getting to their questions?

On a bright October day in 1975 a strange hat-shaped craft came down through the clouds. As it penetrated the atmosphere the incredible, computerized robot released three parachutes to slow its descent. It landed without damage on a rocky plain and then, for 53 minutes, the alien craft sent back pictures and reports to a group of scientists on another planet.

The planet on which it landed was Venus and the scientists, of course, were Earthmen. They monitored the signals for nearly an hour before the intense heat and crushing atmospheric pressure of Venus stopped the transmission. But the arrival of the Soviet Union's Venera 9 on Venus was—despite the short period of operation—a momentous, sensational event. For the first time man glimpsed the surface of Earth's nearest planetary neighbor whose face is permanently shrouded in cloud. Immediately, a host of theories had to be discarded because the strange world was not what had been expected. A few days later a second probe, Venera 10, landed on another part of the planet and sent back very different pictures for 65 minutes.

Within eight months the United States was equally successful in landing two complex craft on the surface of another planet—Mars. This time, because the temperature is not as high as on Venus and the pressure is even less than that on Earth, Vikings 1 and 2 were able to transmit television pictures and information for long periods, enabling scientists to carry out many experiments. Not only did the Vikings show us the Martian terrain but

Opposite: planet Earth photographed by Apollo 11 astronauts on their way to the Moon in July 1969. North Africa and Arabia are clearly visible through gaps in the cloud. Now that we have begun to explore our Solar System in manned and unmanned craft, many of the questions that have puzzled astronomers about the other planets may be answered.

The Scale of the Solar System

they also scooped up soil and examined it in a miniature laboratory for signs of life.

The Mars and Venus space missions contributed more to our understanding of those planets in a few days than centuries of telescopic observations. But they are just part of an elaborate, ambitious program to explore the entire enigmatic Solar System in an attempt to provide answers to age-old mysteries. Spacecraft have already flown close to Mercury and Jupiter, and future projects will send more advanced probes back to these planets and on to Saturn, Uranus, and Neptune in the 1980s.

As with every other area of astronomical research, however, it seems that the more we learn the more puzzling the cosmos becomes. It is a sobering thought that while some astrophysicists are confident they know the origin and structure of the Universe, many aspects of the planetary system of which we are a member

Right: a view of the Solar System with the Sun in the center. The nine planets all move around the Sun in their own paths and each takes a different time to complete its orbit around the Sun. The planets are not drawn to scale because the outer planets are in fact much farther from the Sun than the near ones and would be impossible to show in scale. The artist shows clearly, however, how the planets have no light of their own and simply reflect the Sun's light.

still mystify us. Astronomers are still not sure how it was formed and whereas distant objects give up their secrets willingly, our closest neighbors continue to puzzle us.

Before examining the new light that space exploration has shed on the Solar System, let us look at the size and scale of this fascinating collection of planets and moons. Imagine that the Sun is the size of an orange. The Earth, third from the Sun, would be the size of a pinhead in comparison and 25 feet away. Jupiter, the largest of the nine planets and fifth from the Sun, would be the size of a large pea 130 feet away. Pluto, the outermost planet, would be even smaller than a pinhead 1200 feet from the system's center. Where would the nearest star, Alpha Centaurus, be in relation to this miniature Solar System? It would be some 3000 miles—approximately the distance that separates Great Britain and the United States.

Below: the Sun and the planets of the Solar System drawn to scale. The planets are shown in order of their relative distance from the Sun. Mercury (far left) is nearest to the Sun. The others, from left to right, are Venus, Earth, Mars, Jupiter (the largest), Saturn, Uranus, Neptune, and Pluto, the farthest away. The broad golden curve at the bottom is a part of the Sun. The difference in size between the four great planets and the five small ones is clearly shown. Jupiter, for example, could hold more than 1300 Earths!

There is an awe-inspiring orderliness about the Solar System. All the major planets orbit the Sun in a counterclockwise direction (when viewed from the North Star) and each rotates about its axis in a counterclockwise fashion, too. Even the Sun rotates in the same direction. The planets have nearly circular orbits and seem to be placed at regularly increasing distances from the Sun, all orbiting roughly in line with the Sun's equator. With just a few minor exceptions the planets' satellites also go around in near-circular orbits in a counterclockwise direction and are also close to the plane of the planetary equator.

These facts, known for many years, suggest that the Solar System was created out of a single event, imparting similar properties to each member of the group. But what could have been responsible? There have been many theories put forward to explain the mechanics of planetary creation, some reasonable,

Origins of the Solar System

Below: sequence of paintings illustrating the "catastrophe" theory of how the Solar System was formed, put forward by Austrian Hans Hörbiger in 1913. A block of cosmic ice crashes into a hot metallic star (1) causing a vast explosion (2). The spinning star spews out a rotating mass of molten matter (3) which eventually cools down to become our Solar System of today.

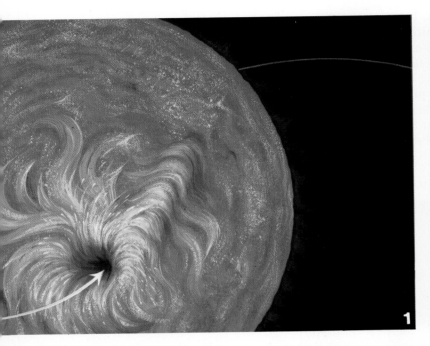

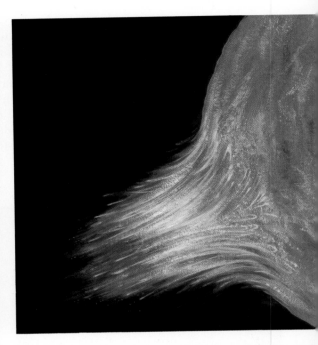

others outlandish. Georges de Buffon, an 18th-century French naturalist, suggested that the planets were formed from debris created when a comet smashed into the Sun. Another "catastrophe" theory, put forward in 1913 by Hans Hörbiger, an Austrian inventor and engineer, suggested that the Universe was made up of hot metallic stars and cosmic ice. When a small star lost its heat it became coated with ice and if it then collided with a large, hot star it became embedded and slowly melted, causing a great explosion. The blast threw off hydrogen and ice and planets condensed out of this material. He believed the planets near to the Sun were ice covered and that Earth was just far enough away to escape that fate. He also argued that moons were thrown off in a similar way by the Sun and that Earth had captured and lost six previous satellites before grabbing our present Moon—which was formerly a planet named Luna! We now know that the Austrian's theories are totally wrong but in the early 1900s he had millions of followers, some of whom broke up learned meetings crying: "Out with astronomical orthodoxy! Give us Hörbiger!"

Some writers and theorists have overcome the planetary evolution problem by suggesting that we live in an "inside-out" world, deep within a hollow Earth. At the center is a dark blue sphere on which points of light move about, creating a "phantom" universe. Like every other theory it had its followers, but a far more plausible theory was put forward by two American astronomers, Chamberlin and Moulton, who suggested in 1905 that the Sun and another star may have suffered a near collision. As a result of the exceptionally strong gravitational attraction that would develop during such a near-miss, a long stream of gaseous matter would have been drawn from one of the stars and this would then break up and condense into planets. It was an attractive theory and one that seemed to fit the facts. Such a stream of gas might well bulge in the middle and taper at each end—explaining why the largest planet, Jupiter, is at the center of the nine planets, and why Mercury and Pluto, the nearest and farthest planets respectively, are so small. If this is what happened, then planets must be a very rare phenomenon because it is estimated that only 10 such near collisions between stars occurs in the lifetime of a galaxy. Calculations have shown, however, that this theory is not correct: The planets would have been thousands of times farther from the Sun if they had been created in such a fashion.

Above: Hans Hörbiger. He died in 1931 but his theories remained popular, most of all in Nazi Germany. We now know that his theory is totally wrong, but during his lifetime he had millions of followers, some of whom broke up learned meetings shouting "Out with astronomical orthodoxy! Give us Hörbiger!"

Below: a theory, proposed by two Americans, J. F. Chamberlin and T. C. Moulton, in 1905 suggests that the planets were formed when another star grazed the Sun in passing. The strong pull of gravity tore a mass of gaseous matter from one of the stars (left) which then broke up and condensed to form the planets (right). This theory, though attractive, now seems improbable, largely because such a "catastrophe" is highly unlikely, but also because it would have resulted in an even wider distribution of the planets.

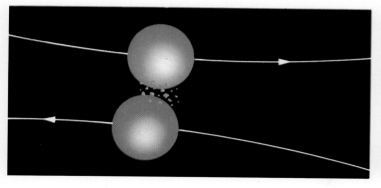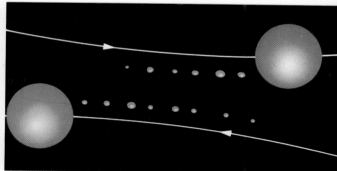

Above: drawings demonstrating a third theory about the origin of the Solar System. This starts from the premise that a contracting cloud of matter would throw off rings of matter as its rotation increased. When the central cloud had become the Sun, it would be surrounded by several rings of matter which would evaporate, cool, and condense to form planets. First put forward in 1796, it was taken up again in 1944 by Karl von Weizsacker a German astronomer. Known as the "gravitational" theory, it still has its followers, though it too fails to explain many of the irregularities in our Solar System.

By 1939 it was generally agreed that matter ejected by the Sun would be so hot that it would expand into a thin gas instead of condensing, so a new theory was needed. Instead, an old one that had been discarded was re-examined. Pierre de Laplace had suggested in 1796 that a contracting cloud of matter would throw off rings as its rotation increased. By the time the matter at the center had turned into a star it would be surrounded by several rings of matter that would, in turn, condense into planets, throwing off their own rings. These, in time, would condense into moons. This idea seemed to fit the facts. In the light of this theory, the asteroid belt between Mars and Jupiter was a ring of matter that had not formed into a planet, and Saturn's rings were also made of matter that had failed to become a moon. But, after being fashionable for years, Laplace's theory was dropped when James Clerk Maxwell showed, in 1859, that it was mathematically impossible for a ring of gaseous matter that had been ejected by a body to form into a planet or a moon. It would condense only into small particles.

When other models of planetary creation had been disproved Laplace's was again re-examined in conjunction with man's new knowledge of galaxies. This led Carl von Weizsäcker, a German astronomer, to suggest in 1944 that vast collections of matter in the Universe would break up into eddies caused by turbulence. As a result, separate systems would condense and further sub-eddies on the outskirts would produce planets. This theory was modified by others to take account of magnetic and gravitational forces, making it far more acceptable to most astronomers. But there remain so many irregularities in our Solar System that no one theory has succeeded in explaining all the phenomena we see. Some astronomers believe they have found another planetary system in our galaxy in the process of forming near the star Epsilon Aurigae. This star is eclipsed every 27 years by an orbiting dark companion that scientists suspect is an embryo star-and-planet system.

Those are the theories. Since we entered the Space Age, however, we have accumulated a large body of knowledge of our Solar System based on observation and study with the most up-to-date equipment. We can build up a picture of our planetary neighbors starting with Mercury, the innermost body, and working outward through the system.

Because it is so close to the Sun, direct observation of Mercury

Mercury - the Scorched Planet

Left: an artist's impression of the landscape of Mercury, the planet nearest to the Sun. Mercury has no protective atmosphere, and its nearness to the Sun causes temperature variations from 430°C during the day to bitter cold during the night. The picture shows a volcanic plug in the foreground, worn down by the alternate expansion and contraction of the planet's surface caused by extremes of temperature. The blazing Sun—three times larger than it appears to us on Earth— is fortunately hidden behind the crumbling mass of lava. It illuminates an inhospitable, featureless landscape. In the sky at the extreme right is Earth and its satellite the Moon, looking like a double star in the black sky.

Below: the surface of Mercury as photographed by the space probe Mariner 10 in 1974. Its strangely lunar landscape, though it could not possibly support life, could be explored more thoroughly by future astronauts, possibly in underground bases.

Venus - the Clouded Planet

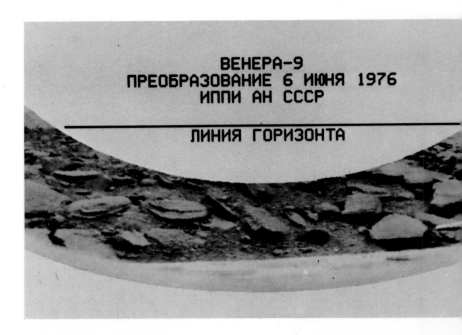

Right: close-up view of the rocky surface of Venus, photographed by the Soviet spacecraft Venera 9 in October 1975. Experts were amazed by this and other photographs because they showed a rugged surface. Most had expected that Venus would be worn smooth by erosion.

Above: the cloud-covered surface of Venus, photographed by the United States Mariner 10 space probe in 1974. Clouds of sulfuric acid permanently cover the planet, and only radar observation from Earth or probes landing on the surface could tell us more about its landscape and other features, such as a surface temperature of about 460°C night and day at the equator and poles.

is difficult. Astronomers believed that it kept the same face turned toward the Sun permanently (as the Moon does to the Earth) but radar measurements have shown that it rotates on its axis once every 59 days. It is small and Moonlike in appearance and America's Mariner 10 spacecraft found a weak magnetic field—much to scientists' surprise—and a *magnetosphere* (field of radiation). It probably has the widest temperature range in the Solar System, with temperatures of 400°C on the sunny side and −200°C on the night side. It is an inhospitable planet and unlikely to support any form of life.

The beautiful, bright "morning star," Venus, was regarded until as recently as the early 1960s by many astronomers to be the most likely candidate for supporting some form of life. It is of similar size and mass to the Earth, but its permanent cloud cover prevented astronomers from seeing and examining its surface. It was thought to be a warm, dark world covered in oceans. But the information sent back by seven Soviet Russian space probes, which descended into the atmosphere, and two American Mariners, which flew past the planet, told scientists it was a far from welcoming world. The two Venera spacecraft that landed and photographed the planet for short periods in October 1975 confirmed that view.

The clouds consist mostly of sulfuric acid, while the main component of the atmosphere is carbon dioxide. The temperature on the surface is around 460°C and its atmospheric pressure is about 90 times that on Earth. Surprisingly, the dark side of the planet is warmer than that facing the Sun and despite the thick cloud cover the illumination on the surface of Venus is high. Another puzzle is the virtual lack of a magnetic field. It is also the flattest of the "terrestrial" group of planets—those similar to Earth. America plans to send two Pioneer spacecraft to Venus in 1978, one of which will launch four entry probes to different parts of the planet's surface. But it will take many years of intense investigation before Venus loses its title of "the most mysterious planet in the Solar System."

Knowledge of the next planet, our own Earth, has grown enormously since we entered the space age. Before then we were hampered by being confined close to its surface but satellites and then astronauts have been able to take a distant view of the planet discovering, in the process, that it is slightly pear-shaped and its equator is elliptical. Probably the biggest surprise has been the discovery of the Van Allen belts of radiation that appear to be made up of charged particles trapped in the Earth's magnetic field. These Belts are now called the magnetosphere. It was thought at first that Earth's magnetic lines of force were symmetrical, like a huge doughnut, but later satellite observations showed that the *solar wind* (fast-traveling particles from the Sun) "squashed" the magnetosphere on the side of the Earth facing the Sun while allowing it to extend far into space on the other side.

Intense research has been carried out to solve the mysteries of our nearest neighbor, the Moon—the largest satellite in the Solar System. It has been the target of 57 individual space missions between 1959 and 1975. It presents the same face to us

Below left: *The Man from Venus*, an illustration from a 1939 science fiction magazine. Apart from the casual way the Earth visitor has taken off his space helmet, the green and watery landscape of the planet and its web-footed inhabitants shown here are probably the artist's idea of what life on a cloud-covered planet would be like.

Venus - Earth's First Colony in Space?

Venus, we now believe, has an atmospheric pressure more than 100 times that of Earth. Its atmosphere is composed largely of carbon dioxide, and the Sun's light is constantly obscured by clouds of sulfuric acid. The temperature reaches 500°C. Yet Venus is the most likely objective of Earth's space colonization plans.

By about 2080, man will have the technology to send spacecraft into orbit around Venus. These craft will fire rockets into the scorching atmosphere—rockets packed with some of Earth's most primitive life forms, blue-green algae. The rockets will then explode before crashing and scatter the algae into the clouds of sulfuric acid.

Scientists believe that, warmed by the Sun, the algae will then begin to consume the carbon dioxide in the Venusian atmosphere, at the same time releasing oxygen as a waste product. As the carbon dioxide is consumed, more heat will escape from the "greenhouse" atmosphere of the planet. As temperatures fall, the water vapor in the clouds will condense and fall as rain! In time, the clouds will roll away and the Sun will shine fully onto Venus at last. Later, the surface should cool sufficiently to support other plant and even animal life transported from Earth. If the scientists are right, Venus could become Earth's first colony in space!

Our Moon Explored

at all times, because of gravitational attraction, so it was not until spacecraft had orbited the dead world that astronomers could map the entire surface. Man first walked on the Moon on July 21, 1969, and subsequent manned flights have produced a wealth of information as well as rock and soil samples. Most of the Moon's scarred face seems to have been formed in the early part of its life, 400,000,000 years ago. Huge craters were created, probably by the impact of great meteorites or comets, and some of these were filled by dark-colored lava flows that created the "seas" that men could see in their telescopes. More recent craters, such as Copernicus and Tycho, are distinguishable by being better preserved than the others, and they, too, seem to have been formed by meteoritic impact. In addition to the craters there are *rilles*, which are canyonlike features sometimes up to 100 miles long, a mile wide, and 1000 feet deep. They may have been lava channels.

Because the Moon has virtually no atmosphere, debris from outer space can crash on to its surface without resistance. As a result the surface has been churned up, causing a deep dust layer. The temperature change, from 120°C to −180°C every two weeks assists the break up of the lunar rocks.

Below: the Apollo 11 lunar module rises from the surface of the Moon to meet the command module for the return flight back to Earth in July 1969. Astronauts had finally escaped from Earth's limits, walked on the Moon's surface, and begun the systematic exploration of space.

Left: the Lunar Rover on the Moon's surface. Used on three Apollo missions, this vehicle enabled United States astronauts to explore more widely and to collect a total of about 660 pounds of lunar rock to bring back to Earth.

Below: scientist examining a sample of lunar rock. Despite 3000 tests of all kinds made by 200 research teams around the world, no signs of life have been found, though we now know more about lunar rocks than we do about Earth's rocks.

Bottom: The surface of the Moon may seem gray and uninteresting, but rock samples sliced into thin sections and examined under a microscope reveal a kaleidoscope of color, as in this astonishing photo-micrograph.

In a way we now know more about lunar rocks than about their counterparts on Earth. About 44 pounds of the samples brought back by Apollo astronauts and the Soviet Union's Luna 16 and 20 robots, were divided between close on 200 research teams around the world. Their results appeared in the reports from the seven Lunar Science Conferences held at Houston, Texas, between 1970–77 and which, combined, run to 30,000 pages. Elaborate quarantine precautions were enforced when the first astronauts returned to Earth for fear that they had brought back deadly organisms that might contaminate life on Earth. But 3000 tests on lunar samples have failed to find any signs of life. Despite all this information, however, there are still many unanswered questions, particularly about the Moon's origin. Was it formed at the same time as the Earth? If so, why are many of its rocks very different and slightly older than those on Earth? Why is the Moon's density not as great as ours? What is the source of the magnetism found in its rocks? What caused the tremendous bombardment of its surface in the early part of its history, creating its pockmarked surface?

One theory suggests that the Moon was once a planet even closer to the Sun than Mercury. It may then have been deflected from its orbit by close encounters with Mercury and then captured by the Earth. If at that time our planet had smaller satellites these might have crashed into the surface of the new Moon causing giant craters. Exploration of the Moon continues in 1980 with a pair of unmanned satellites from the United States, one of which will take up a pole-to-pole orbit to provide the first global survey of a body other than Earth. These probes may answer some of the questions—but they may find new mysteries, too.

Ten space missions were made to the "Red Planet," Mars, starting in 1965, before the historic Viking 1 and 2 landings in

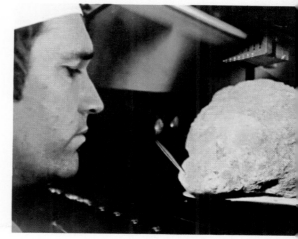

The Red Planet

Above: a recent photograph of Mars taken through a telescope. The distinct dark markings appear to change in shape from season to season. The polar ice cap is clearly visible at top right.

Opposite, top: an imaginative 19th-century artist's view of the Martian landscape. Mars has long been the focus for the most enthusiastic speculation about life elsewhere in our Solar System. When Italian astronomer Giovanni V. Schiaparelli observed canals on the surface in the 1870s many were convinced that this proved there must also be intelligent life on Mars.

Opposite, below: conclusive proof that Mars is in fact red is provided by this first color photograph transmitted from unmanned probe Viking 1 in July 1976. Similarly, the long-held belief that Mars was once as habitable as our Earth was also finally confirmed when the probe signaled the presence of two gases—nitrogen and argon, both necessary basic ingredients of our atmosphere. At least one scientist, Michael McElroy from Britain, gave as his belief that it was "difficult to escape the conclusion that the Martian atmosphere was once very much like the Earth's."

1976, which were principally designed to answer the age-old question: Is there life on Mars? The answer, after many fascinating tests, is that we still do not know. The Viking invasion resulted in superb color pictures of the planet (confirming that it is indeed red) and they took soil samples for instant laboratory tests. Water was mixed with the red dust to bring to life any dormant microorganisms that might be present. The first results seemed to indicate an organic presence but other tests were negative. Was the Viking equipment malfunctioning? Viking 2, which landed on another part of the planet, carried out the same experiments with the same results. Earth scientists are still trying to understand the cause of this remarkable and unexpected reaction. As one put it: "We didn't find alien life but we did find alien soil."

Martian geology has also puzzled scientists. The boulder-strewn scene from Viking 2, said one of them, was contrived almost, "like a movie set painted by someone who doesn't know anything about geology." Man does not yet understand the processes that brought about this startling result. The current theory is that Mars once had a sizeable atmosphere dominated by nitrogen, probably 4500,000,000 years ago. It may have supported rain because although it is now dry on the surface there is evidence of thousands of channels (but no sign of the canals for which the planet was once famous). Huge banks of cloud have been seen in the Northern Hemisphere and freezing fog has been detected in some of the craters. There may be a lot of water still locked up in the planet, perhaps frozen beneath the surface.

The Martian atmosphere is much thinner than scientists had expected and consists mostly of carbon dioxide. Winds up to 300 mph sometimes lash the planet causing great dust storms that obscure much of the surface and create changing features (as seen from Earth) that scientists once thought were signs of growing vegetation. There are four very large features that look like volcanoes but the lack of seismic activity means another explanation may have to be found. There are also many rilles that look remarkably like Earth's canyons. One, known as the "Valley of the Mariners," is gigantic: almost three miles deep, 50 miles wide, and long enough to stretch right across the United States. A lot of water would be required to form such a canyon. But astronomers have noted that they are remarkably similar to the smaller rilles found on the Moon where it seems there has never been any water. So what caused them?

One theory is that comets may have provided Mars with water or gas for brief periods of its history—and there is certainly evidence that it has undergone several changes of climate—in which case, even if there is no life now, future probes may uncover the remains of long-dead organisms that once thrived on the planet.

The giant planet Jupiter is large enough to contain all the other planets in the Solar System. It seems to be a largely gaseous body whose outer temperature is extremely cold. Pioneer 10 was the first probe to pass the planet, in December 1973, sending back pictures and data that changed nearly all our scientific ideas about Jupiter. It has a much stronger magnetic field than Earth and an

Jupiter - the Giant Planet

Below: a telescopic view of Jupiter, the giant planet which is by volume 1000 times bigger than our Earth and twice as immense as all the other planets put together. The "Great Red Spot," Jupiter's most unusual feature, seems to be a semipermanent solid body that changes shape as it moves among the planet's clearly visible belts of cloud.

enormous region of trapped electrons and protons, making it a very dangerous place to visit. The space probe did not behave as expected when it reached the planet and the second Pioneer was reprogrammed to fly over Jupiter's pole so as not to experience so much interference from the massive, magnetic body. Two years and two months after its close encounter with Jupiter, Pioneer 10 was speeding on out of our Solar System when it suddenly came under Jupiter's influence once more. It was beyond the orbit of Saturn when its instruments for measuring solar wind suddenly registered zero for a 24-hour period in mid-March 1976. The solar wind had not stopped; it had simply been deflected by the planet's enormous magnetic tail for at that moment the Sun, Jupiter, and Pioneer 10 were all in line. We now know that this long extension of its invisible magnetosphere stretches for more than 400,000,000 miles, which means that Saturn passes through it once every 20 years producing, it is thought, some unique planetary phenomena.

One particular feature of Jupiter, the Great Red Spot, has

puzzled astronomers for more than three centuries, though it did not flare up and become prominent until 1878. It seems to be a semipermanent, solid body that changes shape and intensity as it floats in the planet's cloud belts. The 1977 Mariner 11 and 12 space probes will photograph the planet from close quarters. A 1982 mission will even enter the planet's atmosphere and when that happens it may encounter more than the Great Red Spot. Carl Sagan, the American astronomer and biologist, and astrophysicist E. E. Salpeter suggested in 1977 that Jupiter's atmosphere supports "abundant biota," in other words, living organisms. These may be huge, floating, balloonlike creatures many miles across. So, even if there is no life on Mars, we may still not be the only inhabited planet in the Solar System.

Another giant planet is Saturn with its beautiful rings. Radar reflections have shown that they are composed of small boulders averaging about 40 inches in size. Three distinct ring systems, the largest being 170,000 miles in diameter, have been known for centuries but in 1976 the existence of a fourth, inner ring was

Left: *Life on Callisto*, a science fiction illustration of 1940. Most scientists now believe that Callisto, one of Jupiter's 13 satellites, may well be composed mainly of ice—an unfriendly and unlikely place even for such monsters as the one shown in this picture. However, some scientists have suggested that Jupiter's own atmosphere may support living organisms, though these may well be huge balloon-like creatures several miles across.

Above: this amazing photograph of the planet Saturn shows its system of rings clearly and fairly open. These rings are Saturn's most distinctive feature, and are believed to consist of billions of tiny particles of dust or ice, packed close together to give the impression of a flat disk round the planet's equator.

Right: Saturn as it would appear from Rhea, one of this planet's 10 satellites. Four other satellites are clearly shown. Seen from Rhea, Saturn's rings seem like a thin white line. The surface of Rhea, it is believed, resembles that of our own Earth's satellite, the Moon. Because this planet, like Jupiter, has a gaseous and intensely cold surface, manned or even unmanned space probes would probably have to land on one of its satellites, such as Rhea, with its lunar landscape.

confirmed. When future Pioneer space probes reach the planet, mission controllers will have to find a way of avoiding the 10-mile thick rings, or guiding the Pioneer space probes through the hoop between the inner rings, to prevent the spacecraft from being punctured by the particles.

Until the beginning of 1977 Saturn's rings were thought to be unique. Then Dr James Elliot and his team at Cornell University, New York, announced that Uranus, too, has rings. They are too faint to detect visually and are far thinner than Saturn's. The discovery was made by plotting the disappearance and reappearance of a star as it was eclipsed by the huge, cold planet. Dr Elliot's team, and two other groups of astronomers, realized from their observations that Uranus had rings of particles, probably rock and ice, similar to but less impressive than Saturn's.

The existence of Neptune, which is very similar to Uranus, was predicted before it was discovered in 1846 to account for deviations in the latter's orbit. Three years later the prediction was confirmed optically.

In the same way, the last of the nine planets, Pluto, was

Saturn – the Ringed Planet

Below: three photographs of Saturn, taken at six-year intervals. As the planet travels around the Sun in its orbit of nearly 30 years, we on Earth see these different aspects of the ring system of what is probably the most beautiful of all planets in the Solar System.

The Outer Planets

Below: William Herschel, the self-taught English astronomer who, in 1781, first discovered the planet Uranus. He was then working on a "survey of the sky" when he saw an object that showed as a disk, and therefore could not be a star. He thought he had found a comet, and named it the Georgian Planet for his king George III.

predicted from deviations in Neptune's orbit, but it took 16 years to find the lonely, small, cold planet. The discovery was made in 1930 but the mystery remains, for it seems to be too small to account for Neptune's orbital disturbance, leading many astronomers to believe that there must be a 10th planet in the Solar System.

As well as the planets there is a belt of asteroids orbiting between Mars and Jupiter. These are sometimes called the minor planets and are thought by some to be either debris from a planet that disintegrated (or whose inhabitants blew it and themselves to pieces), or the "building bricks" of a planet that never formed. They occupy a large gap in the Solar System that would comfortably take a planet. Occasionally an asteroid

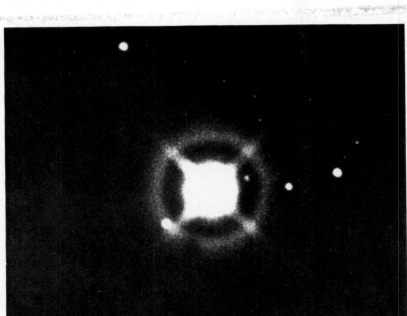

Above, right: Uranus with its five satellites. The ring of light is a purely photographic effect. The faintest and smallest of the satellites, Miranda, orbits at a distance of only 74,000 miles from the planet.

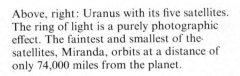

Right: Neptune, first discovered in 1846, is slightly larger than Uranus, but is so far out in the System (2,800,000,000 miles from the Sun) that it is invisible to the naked eye and visible only in outline by means of a telescope. Neptune has two satellites: Triton, which is nearly as big as Mercury, and Nereid (arrowed) which has a distant and erratic orbit.

Left: artist's impression of the surface of Pluto, smallest of the distant planets. Smaller than either Earth or Mars, Pluto, even if it ever had an atmosphere must now have none. It must be either frozen in the rocks or liquefied. Here the artist shows an icy cave from which we see a placid lake of liquid methane. The Sun, tiny at Pluto's distance in the System, shines starlike in the center of the painting. Pluto must surely be the least accessible, the least hospitable, of all the planets.

Below: the two photographs taken at Lowell Observatory, Arizona, in 1930 by Clyde Tombaugh which convinced him and other astronomers of the existence of Pluto (arrowed).

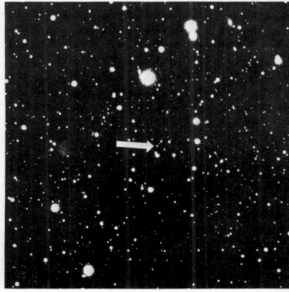

strays dangerously close to Earth, such as the one that passed within 750,000 miles—three times the distance of the Moon—in October 1976. Its diameter was only a few hundred yards and it was the smallest object ever observed in space by astronomers, but it could have caused a disaster if it had collided with our planet.

Even more mysterious are the comets that put in occasional, spectacular appearances. Although they travel deep into space they are members of our Solar System, orbiting the Sun in great, elongated ellipses. There is much argument about their nature and origin but it is thought that a comet's nucleus is less massive than an asteroid, with a long tail of dust and gas. Their origin is puzzling but their orbits have led some astronomers to suggest that a swarm of cometary bodies was once formed in the position now occupied by Jupiter, and were ejected when that planet formed. Most left the Solar System but one in 40 went into orbit. Another theory is that there is a reservoir of comets orbiting the Sun at a vast distance, some of which are dislodged and sent into orbit by passing stars. The mystery may be solved in 1986 when the famous Halley's Comet comes in to view again. Edmond Halley correctly predicted the appearance of this comet in 1759 (though he did not live to see it) on the basis of its 75-year orbit,

Below: Part of the famous Bayeux Tapestry showing a "strange star" in the heavens as seen in 1066 A.D. In 1682 the British astronomer Edmond Halley, for whom it is now named, observed the same comet in the sky and predicted its return every 76 years. So far his prediction has proved correct, and the next sighting is due in 1986.

which takes it well beyond Neptune. American scientists hope to build an unmanned craft with an enormous sail that will use the solar wind to chase and inspect the comet on its next visit.

One man who is sure that comets play an important role in the Solar System is Dr Immanuel Velikovsky, Russian-born scholar, scientist, and author of books about his theory, including *Worlds in Collision* and *Ages in Chaos*. Velikovsky believes that in comparatively recent times Jupiter ejected a massive comet which, on its periodic return to the center of the system, came very close to Earth and Mars. Its first appearance would have been about 3000 B.C. and its subsequent approaches and the havoc it caused led to the worldwide myths and legends of catastrophes and fights between the gods in the heavens. Velikovsky argues that such stories as fire and hail raining down from the sky, the plagues in Egypt that caused skin irritation, the Nile and other rivers turning to "blood," the Sun standing still in the sky and then changing direction, can all be attributed to the effects of this great comet's tail as it swept past the Earth and changed our planet's axis and orbit. Mars was affected, too. Eventually the comet settled down in an orbit around the sun— and became known as the planet Venus. It sounded preposterous

Halley's Comet

Below: when Halley's Comet makes its next Earth orbit in 1986, United States scientists plan to build a vast unmanned spacecraft, driven by solar wind, to chase and inspect the comet at the closest possible range. One problem is that the comet has an orbit opposite in direction to that of the Earth and other planets. This makes the sending of a probe complicated from a navigational and guidance point of view.

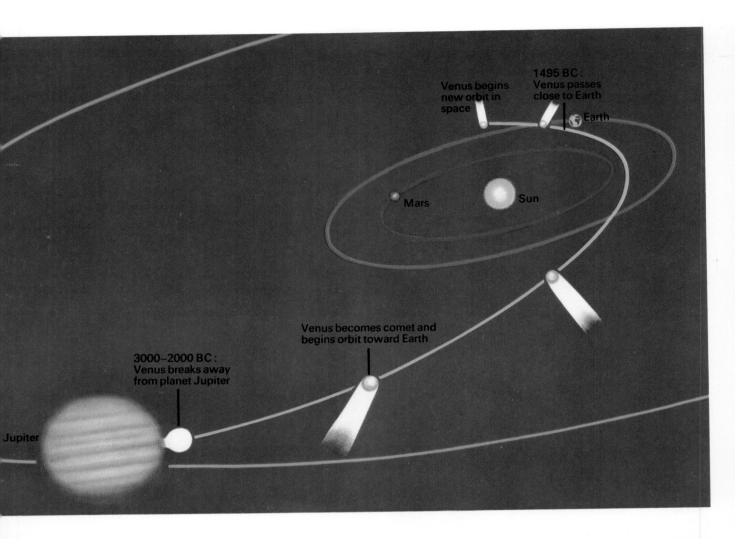

Venus begins new orbit in space

1495 BC: Venus passes close to Earth

Earth

Mars

Sun

Venus becomes comet and begins orbit toward Earth

3000–2000 BC: Venus breaks away from planet Jupiter

Jupiter

Above: these diagrams demonstrate the theory of Velikovsky about planets that collide in space and their effects on the history of the Earth. It all starts with Venus breaking away from Jupiter, for which Velikovsky finds his evidence in the Greek myths. According to the Greeks, Athene—who is always associated with the planet Venus—was born from the head of Zeus, associated with the planet Jupiter. Later, as Venus careered around space before settling into an orbit, it nearly struck Earth. It was at this time, according to Velikovsky's account, that Egypt was afflicted with the plagues recorded in the Old Testament. These are said to include the shower of vermin and three days of darkness.

to astronomers, despite the impressive evidence that Velikovsky had amassed from many sources, but the public found the theory attractive and his books became best sellers. Then, incredibly, the space probes appeared to confirm some of his predictions. Venus, in particular, is much hotter than scientists expected. They now say its clouds have created a "greenhouse" effect, but Velikovksy's supporters believe it is hot because the new planet is still in the process of cooling. It certainly seems to be a younger and flatter planet than Mars or Earth. Another puzzle is why Venus always shows the same face to Earth at its closest approach. Is it evidence that the two planets were once affected by a strong gravitational attraction? Its rotation is different to other planets, too, indicating perhaps that its origin is not the same.

Velikovsky predicted that the Earth had a magnetosphere that was much stronger and more extensive than anyone had expected, and that Jupiter would be a source of radio emissions. These forecasts, and those about Venus, were published in 1950 and have now been confirmed. Is it coincidence or is Velikovsky right? His theory may in time be disproved but at the moment there are too many mysteries in our Solar System to be sure. Besides, as Dr Lynn E. Rose, professor of philosophy at the

Below: in the 8th century B.C., says Velikovsky, Venus came close to colliding with Mars (1). The shocks are recorded on Earth as encounters between the gods. In the 7th century B.C., Mars came especially close to Earth (2): in 687 B.C. an Assyrian army was wiped out by what the Bible calls "an angel." A meteoric cloud? Later (3) Venus and Mars took their present orbits.

The Theory of Velikovsky

Left: Dr. Immanuel Velikovsky, the scientist, scholar, and author of a challenging theory of the part that the formation of the Solar System as we know it today played in early man's mythology and history. He studied as a psychoanalyst under Freud, and in the course of researching some of Freud's ideas, was drawn into a study of Moses and the Exodus from Egypt. He became convinced that the biblical story did in fact record great cosmic events, and has spent his time since then writing about his theory in books of great and impressive scholarship. These impress some, but infuriate other more traditional fellow scholars.

Are We Being Watched?

State University of New York, Buffalo, has observed, Velikovksy's is "just about the only theory of the Solar System and of ancient history that has *not* had to be drastically revised during the past two decades."

Meanwhile, our Solar System provides us with many other puzzles, and not all of them are related to the origin and nature of the planets, comets, and asteroids. Just occasionally something happens that sends cold shivers up and down the spines of those with enough imagination to believe that someone out there may be watching us.

For years before man reached the Moon astronomers were reporting strange geometrical patterns on its surface, lights inside craters, huge bridges that mysteriously disappeared and a host of other "activities." Were they mistaken or was an alien race using our Moon as a base for watching the Earth? When man landed on the Moon he found it to be, as expected, a barren world. Life could not have developed there nor was there any evidence that other living creatures had paid a visit, either. But what was it that Neil Armstrong—the first man to walk on the Moon—saw shortly after he made his historic leap for mankind onto the surface? According to Otto Binder, a former NASA space program member, significant portions of Apollo 12's conversations with mission control have not been made public. One dialogue, he says, was heard by people using VHF receivers to bypass NASA broadcast outlets. It was as follows:

Mission Control: "What the hell was it?" "What's there?" . . . "Malfunction?"

Armstrong: "These babies are huge, sir . . . enormous . . . Oh God, you wouldn't believe it. I'm telling you there are other spacecraft out there . . . lined up on the far side of the crater edge . . . they're on the Moon watching us . . ."

Was it an optical illusion or did Armstrong see evidence of extraterrestrial visitors? NASA made no comment on the report. Astronauts on Apollo 16 and 17 missions reported seeing strange flashes of light over the Moon's horizon. NASA could offer no explanation. Other spacemen, on flights around the Earth and to the Moon, have reported mysterious objects following them and lights in the sky. Were man's first steps into space being monitored by creatures from another part of the Universe? And why did a transmitter left behind on the Moon by Apollo 14 suddenly stop on January 18, 1976 (five years after going into operation) and then start up again a month later? Why, too, did it function better than ever after its month "off the air" and why did one experiment that had not functioned previously start working perfectly after the inexplicable shutoff? The station has since behaved erratically, so perhaps severe lunar temperatures are to blame.

Unless, of course, "someone" was taking a closer look at the equipment used by the new race of space explorers: the Earthmen. Absurd? Only if we choose to ignore the startling evidence of unidentified flying objects that have now been reported in the skies of our planet for over three decades. It is the most widespread and baffling phenomenon of the century and no examination of our strange Universe would be complete without tackling the great flying saucer mystery.

Opposite: an artist's impression of an alien radio telescope, paraboloid in its design, somewhere in another Solar System. The planet on which it is sited, like our Earth, has a satellite Moon clearly shown at top right, while its Sun sinks redly below the horizon. Perhaps the scientists working at the telescope are observing our own Solar System. Perhaps one day they may even make contact with us!

Chapter 6
The Coming of the Saucers

Are there such things as flying saucers? Reports from as long ago as 1878 indicate that there have been sightings of strange objects in the skies. But from 1947 on there appears to have been a rash of reports by all kinds of people of peculiar saucer-shaped objects shooting across the sky at unbelievable speeds. Here are some of the fascinating cases of observation of curious aircraft—many of them by trained pilots and aeronautical personnel. The standard of evidence varies from one sighting to another, but the question remains: can all these people be imagining things, or are there really unidentified objects sweeping across our horizons?

High above Kentucky the plane climbed at top speed in an effort to close on an unidentified object in the sky. It had been spotted by many observers below, including the control tower operators at Godman Air Force Base. Three F-51 Mustang fighters had been diverted from a training exercise in an attempt to identify the mysterious craft. Captain Thomas Mantell, flying the leading plane, had outdistanced his wingmen when his first dramatic radio call came to the tower: "I see something above and ahead of me, and I'm still climbing." Mantell was asked to describe what he could see. "It looks metallic and is tremendous in size...." Seconds later he called, "It's above me, and I'm gaining on it. I'm going to 20,000 feet." Those were the last words he spoke. Later on that fateful day of January 7, 1948, Mantell's body was found near the remains of his disintegrated aircraft, 90 miles from Godman Base. The unknown flying object had escaped. Some people believed that Mantell and his plane had been attacked by an alien spaceship, and rumors even spread that the dead pilot's body had been burned by a mysterious ray. The United States Air Force explanation did little to soothe growing public uneasiness about flying saucers. According to the version of the story as told by the United States Air Force, the skilled veteran pilot, known to be level-headed and careful, had gone to his death chasing the planet Venus. It seemed incredible to most people that the experienced Mantell and the trained control tower observers could mistake a familiar heavenly body for an object described by some witnesses as "a huge ice cream cone topped with red."

Opposite: this photograph of a flying saucer is probably the most famous of all. George Adamski, who took it in California in 1952, identified it as a "scout ship" from Venus. Was Adamski right? Are the strange objects—some shaped like saucers, others like cigars or upturned bowls— really vehicles from which alien beings are observing life on Earth?

He Called Them "Flying Saucers"

Fifteen days after Mantell's death the USAF's official investigation of flying saucers—now more commonly referred to as UFOs or unidentified flying objects—went into operation. It was code named Project Sign, and its existence indicated that the authorities were in reality concerned about the strange aerial intruders that had flown into the headlines a little over six months before the Mantell incident.

On June 24, 1947, Kenneth Arnold saw nine shining disks weaving their way in formation among the peaks of the Cascade mountains of Washington state, and first alerted the world to the possibility that planet Earth was being visited by beings from another civilization. Arnold—a private pilot, member of the Idaho Search and Rescue Mercy Flyers, and Flying Deputy for the Ada County Aerial Posse—at the time was searching for a commercial plane that had crashed in the mountains. He was after a $5000 reward for locating the lost plane. Arnold never found the wreckage, but his sighting made him world famous.

Describing the flying objects he had seen to reporters, Arnold said that they moved "like a saucer skipping across water." An inventive reporter coined the phrase "flying saucer," and it soon became a household term—even flying into the pages of the dictionary. The popular term is not used by serious investigators, however. They prefer the more exact term UFO.

Kenneth Arnold was not the first man to see strange flying objects in the sky, nor even the first to describe them as saucer-like. A Texas farmer once saw a fast moving object in the sky and said that the only way he could describe it was as a large saucer. That was on January 24, 1878! Strange aircraft were also reported throughout World War II, and were often thought to be secret weapons of the enemy. But Arnold's sighting seemed to mark a turning point in UFO activity as well as supplying the media with a neat and colorful description. Perhaps it was because people recovering from a second major global war in only 25 years were hoping that there were wiser creatures in the Universe who could help put the Earth's troubles into perspective. Perhaps it was Arnold's estimate of the flying saucers' amazing speed of an incredible 1700 miles per hour that caught the public imagination. It might even simply have been the description of the flying crafts' shape that interested and amused people.

Four days after Arnold's saucer sighting two pilots and two intelligence officers saw a bright light performing "impossible maneuvers" over Maxwell Air Force Base in Montgomery, Alabama. On the same day an Air Force jet pilot saw a formation of five or six unconventional aircraft near Lake Meade, Nevada. Another formation of flying saucers was seen by the pilot and copilot of a commercial DC-3 on July 4. They watched the five objects silhouetted against the sunset for 45 minutes. Soon after this group disappeared, a second formation of four came into view. Could they have been the same nine disks Arnold had seen? Only a few days later in early July a supersecret Air Force test center in the Mojave desert of California was being treated to a whole series of aerial visitations. The first sighting was by a test pilot at Muroc, now Edwards, Air Base. He was preparing for a flight in an experimental aircraft when he saw a yellow spherical

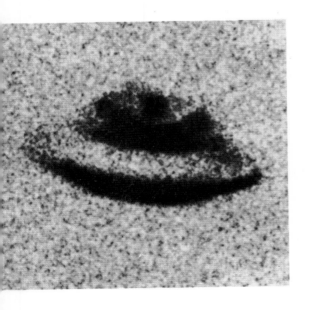

Above: this disk-shaped object was seen and photographed by Gordon Faulkner, a young Briton, over southern England in 1965. Even today, the most common UFO sightings are of saucer or disk-type objects. Many reports mention flashing lights: some are even more bizarre.

Left: Kenneth Arnold, a business man from Boise, Idaho. He made the first widely publicized sighting of unidentified flying objects in June 1947. The term "flying saucer" was coined by a reporter when Arnold said that the movement of the objects reminded him of how a saucer would look if skimmed across water.

craft moving against the wind. Several other officers at the base had seen three similar UFOs ten minutes earlier. Two hours later technicians at nearby Rogers Dry Lake saw a round object of aluminum appearance. After watching it for 90 seconds they decided that the strange craft was "man-made." Later that same day, a jet pilot flying 40 miles south of Edwards saw above him a flat object that reflected light. He tried to investigate but his aircraft could not climb as high as the UFO.

In the early days of UFO reports many people believed that Arnold and others had observed a new secret aircraft or guided missile being developed by the United States military. Arnold himself believed that for a time. But following a spate of sightings from reliable and competent witnesses—many of them experienced pilots fully aware of the latest aeronautical developments—the Air Force got jittery about UFOs. Behind the scenes

The Early Evidence

the USAF was exploring the possibility that they belonged to a foreign power. After all, at the end of the war the Allies had obtained complete data on the latest German aircraft and missiles that had been under development. Had the Soviet Union uncovered a new aerodynamic concept from the German data? Or had they developed their own unique form of aircraft with which to spy on the United States?

It was soon clear that the Soviet Union could not have produced an aircraft that matched the flight characteristics of the UFOs in such a short period. Even if they had been able to do so, it was extremely unlikely that the Soviets would risk exposing the secret so soon by flying over so many countries. For, ever since Arnold had reported flying saucers skimming around the snow-clad mountain peak of Mount Rainier, UFOs had made appearances over every country in the world.

Once it was apparent that no nation on earth was capable of producing the UFOs, the possibility that another civilization was watching our planet had to be faced. This idea had already been put forward by some writers. But there was an alternative. Perhaps "saucer fever" had gripped the nation, causing ordinarily cautious people—and even experienced observers—to see UFOs in everyday objects. This seemed to be the official line as report after sensational report was explained away as balloons, birds, meteorites, planets, conventional aircraft, or natural phenomena. So why did investigations under Project Sign continue, to be followed in time by Project Grudge and Project Blue Book? Because a small but significant percentage of UFOs reported remained unidentified and unexplained as natural phenomena after experts had ruled out every known possibility. Often these were the best reports in the files in terms of the reliability of witnesses or the standard of supporting evidence

In the summer of 1947, shortly after Arnold's original sighting, UFO reports flowed in from all over the country as people scanned the sky eagerly and apprehensively.
Right: United Airlines pilot E. J. Smith saw a formation of five UFOs during a trip on July 4, 1947. His copilot verified the sighting. Smith is explaining that the plate he holds shows the approximate size of the disks as they looked to him from the plane.

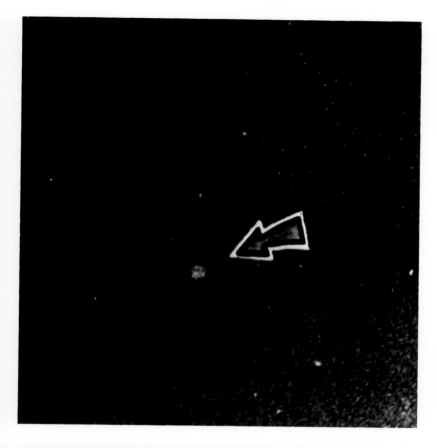

Left: Frank Ryman, a Coast Guard member, managed to photograph a "flying disk" he sighted near Seattle, Washington, in July 1947. The photo is here shown enlarged 20 times, with the UFO indicated by an arrow. This was the very first photograph ever taken of a UFO. The Air Force claimed it was clearly a picture of a weather balloon.

Below: this photograph shows two of the three luminous objects that flashed across the skies of Kentucky in July 1947. Many observers of the glowing lights phoned the Louisville Weather Bureau with their reports, most of which described the lights as flying disks or saucers.

The UFO that Killed

At 1:15 p.m. on January 7, 1948 the control tower of Godman Air Force Base in Kentucky got a phone call from the State Highway Patrol. They said that townspeople about 80 miles away had reported seeing a strange aircraft in the sky. Godman personnel checked and informed the police that there were no flights in the area. Not long afterward, however, the control tower saw the object, which no one could identify.

About an hour later, three F-51 Mustang fighters on a training exercise came into view. The Base commander asked Captain Thomas Mantell, the leader, to investigate the unknown flying object.

All the Mustang pilots knew that to go beyond 15,000 feet without oxygen was dangerous. Yet at 2:45 p.m. Mantell told the control tower that he was going to climb higher to get closer to the odd craft. When the others reached 15,000 feet, they could not contact Mantell, who was above them.

At 3 p.m. the control tower lost sight of the weird aircraft. A few minutes after that Mantell's plane dived, exploding in midair. The wreckage was found 90 miles away. The official explanation of Mantell's death was that he had been chasing Venus. Few believed this implausible reason—and the mystery remains.

such as radar reports.

The military establishment worried that a foreign power could use the flying saucer craze as a cover for spying over United States territory, because the deluge of UFO reports that accumulated during the late 1940s caused an understandable public apathy about objects in the sky. Not that people were no longer concerned, but in most cases it had become physically impossible to do more than log in reports in the hope that some pattern would emerge from the data to aid ultimate identification. Yet until the origin of UFOs was known, they had to be regarded as a potential threat.

After six months of intensive saucer activity, even the most nervous citizens had to admit that if the saucers were hostile they were taking a long time to show it. Then, after the devastating news of Captain Mantell's death, public disquiet increased. At the official level, Project Sign investigators learned that same year of a pilot's duel with a flying saucer. It happened near Fargo, North Dakota on October 1, 1948. Lieutenant George F. Gorman of the North Dakota Air National Guard, was nearing the air base after a cross-country flight. When he had radioed for landing instructions, the control tower had told him that a Piper Cub was the only other aircraft in the vicinity, and Gorman could see the light aircraft below him as he made his approach. Suddenly a light passed him to the right. He thought it was the tailplane light of another aircraft, and called the control tower to complain. When he was told there was nothing else in the sky except the Piper Cub he decided to pursue the light.

As he turned to intercept he could see the other aircraft clearly silhouetted against the city lights. The mysterious light, however, did not appear to be attached to a solid body. Gorman closed in at maximum speed. At 1000 yards he could see that the light was six to eight inches in diameter and blinking. When he got this close, the light became steady and the aircraft banked sharply to the left with his jet on its tail. After a series of abrupt maneuvers Gorman found himself on a collision course with the UFO. When it became clear that the light was not going to take evasive action the National Guard pilot took a dive. The saucer missed the plane's canopy with only feet to spare. The "duel of death" continued, and Gorman had to dive a second time to avoid a head-on collision. Moments later the strange light began to climb. It disappeared as a shaken Gorman returned to Fargo to file his astonishing report.

His account to the Air Technical Intelligence Center (ATIC) included these words: "I had the distinct impression that its maneuvers were controlled by thought or reason." Four other observers saw a fast moving light in Fargo at the same time, but no one witnessed the dogfight.

To saucer advocates, the Mantell and Gorman cases were sufficient proof that superior beings in vehicles of advanced technology were patrolling our skies. Some assumed they were hostile, or at least capable of retaliating if pursued. Others thought the Mantell incident was an accident, and that the Gorman case was no more than a playful game on the part of the spacemen. Meanwhile the Air Force maintained the line of dismissing reports publicly while investigating them with renewed

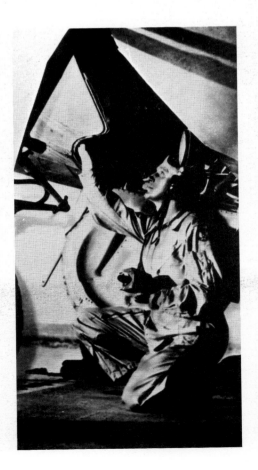

Below: a movie camera being fitted into an F-51 fighter plane of the kind piloted by Air Force Captain Thomas Mantell at the time he was mysteriously killed chasing an unidentified flying object. The cameras were installed about six months after Mantell's death to enable pilots to photograph any strange objects they might encounter.

Toward an Explanation

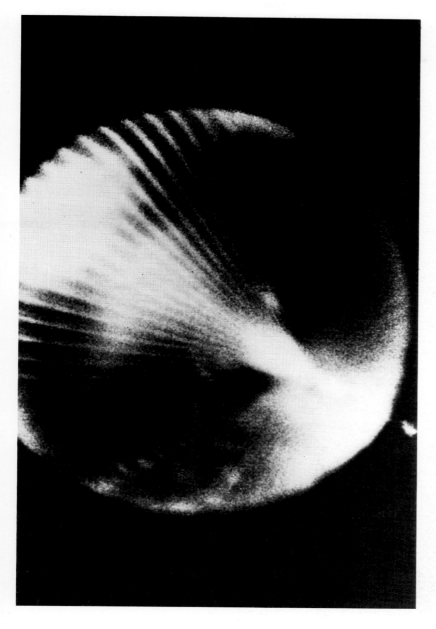

Left: a photograph of a skyhook balloon taken through a refracting telescope. At the time of Mantell's death, information on these secret balloons, designed to collect information from the fringes of the atmosphere, was highly classified. Mantell and the other pilots would not have known anything about these balloons. Official investigators claimed that such a balloon was what Mantell chased to his death.

intensity. Their reasoning became clearer with the publication of the late Captain Edward J. Ruppelt's book *The Report on Unidentified Flying Objects*, published in 1956. Ruppelt had been chief of the USAF's Project Blue Book, successor to Project Sign, from 1951 to 1953. Ruppelt tried to throw new light on some of the famous early accounts that had remained unsatisfactorily explained.

The Mantell case had nothing to do with the planet Venus, he admitted. Ruppelt had discovered that Venus could not have been visible at the time of the chase, though if it were it would have been in the same place in the sky as the UFO. Ruppelt offered a far more plausible explanation. Mantell died while trying to intercept a secret skyhook balloon, he said. The skyhook project, designed to collect information from the fringes of the atmosphere by balloon, was highly classified when Mantell tried

Not UFOs but Skyhook Balloons

Below: Dr Charles Moore, a meteorologist. He observed a UFO with four Navy enlisted men as the group was busy launching and tracking a weather balloon as part of the preparation for a skyhook balloon launch.

to identify the UFO. Ruppelt claimed that everyone involved in the official investigation of the Mantell case was convinced that he had been after a spaceship, so no one bothered to "buck the red tape of security to get data on skyhook flights." By the time Ruppelt checked on this possibility there were no longer any records of the 1948 skyhook flights, but those involved thought that at that time they were being launched from Clinton County Air Force Base in southern Ohio. Winds records for the day in question showed that a skyhook launched from Clinton would have been visible from all the points in the area of Godman Base from which the UFO was seen. The skyhook would have climbed to 60,000 feet, and then drifted in a southerly direction. Mantell died because he did not know he was chasing a balloon, Ruppelt argued. "He had never heard of a huge 100-foot diameter balloon, let alone seen one," he explained. If a skyhook balloon had been launched from Clinton on January 7, even the staunchest believers in flying saucers would have to admit that the facts and description of the sightings tally beyond doubt. If, however, it could be proved that no skyhook balloons climbed into the jet stream over the Godman area that day, then even the saucer skeptics would have to admit that the object which took Captain Thomas Mantell to his death remains one of the most perplexing puzzles of the UFO enigma.

Ruppelt offered an equally plausible explanation for Gorman's strange duel with a flying saucer. Gorman was fighting a small lighted balloon, he said in his book. An almost identical incident had occurred over Cuba in September 1952. A Navy pilot, who spotted a large orange light to the east after making practice passes for night fighters, was told that no other aircraft were in the vicinity. He tried to intercept the flying object. At closer quarters he described it as having a green tail between five and 10 times its diameter. It seemed to be climbing, changing course, and even responding to his plane's movements. After reaching 35,000 feet, the light began a rapid descent. The pilot took two runs at it on a 90-degree collision course. On both occasions the object traveled across his bow. On the third run he reported that he came so close to the light that it blanked out the airfield below him. Then it dived, and although the pilot followed, he lost the light at 1500 feet.

"In *this* incident," Ruppelt wrote, "the UFO *was* a balloon." This was proved by sending a lighted balloon up the following night and ordering the pilot to compare his experiences. "He duplicated his dogfight—illusions and all . . . Gorman fought a lighted balloon too. An analysis of the sighting by the Air Weather Service sent to ATIC in a letter dated January 24, 1949, proved it."

Did it? Major Donald E. Keyhoe, a staunch advocate of the interplanetary spaceships theory and a well-known author of UFO books, agrees that a weather balloon was released in Fargo on the night of Gorman's strange encounter. But he maintains that the Weather Bureau observer who tracked it recorded a course that took it away from the area in which the aerial duel was fought.

Even if the Mantell and Gorman cases were misidentification of different types of balloons, the fact remains that the U.S. Air

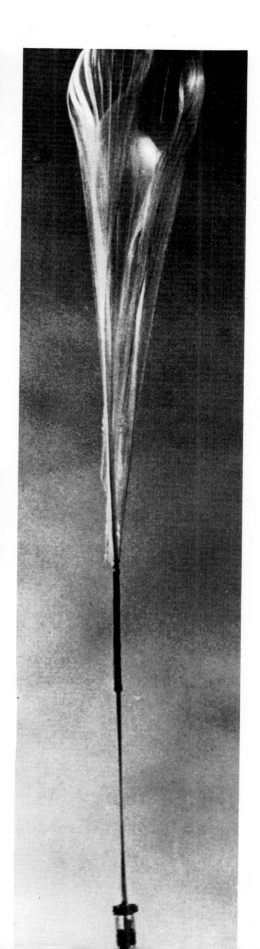

Above: a picture of a skyhook balloon being inflated, one of a group released by the Navy in an attempt to explain the rash of UFOs.
Right: a skyhook balloon in flight. The balloons were 100 feet long, and moved at up to 200 miles per hour when swept along by winds.
Below: a flying saucer reported over London. This UFO turned out to be a high-level cosmic radiation balloon released by a university.

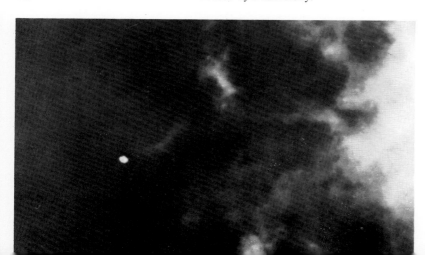

Scientists Join UFO Observers

Below: Dr Lincoln La Paz, an authority on meteorites. When head of the University of New Mexico's Institute of Meteoritics, he became involved in the odd rash of reports of green fireballs in the skies of New Mexico in 1948 and 1949. Originally called in to help discover what the fireballs were, Dr La Paz later sighted a fireball himself. He was certain it was not a meteor.

Force had many more startling cases in its files about which the public knew nothing. An ATIC study of the flying saucer reports received in the year after the Kenneth Arnold sighting had culminated in an historic, top secret "estimate of the situation," which concluded that UFOs were interplanetary. Air Force Chief of Staff General Hoyt S. Vandenberg was not convinced, however, and rejected the report.

Whether people believed in them or not made no difference to the flying saucers. They continued to appear with startling regularity, often giving ground observers spectacular displays of aerial acrobats.

As officialdom found new explanations for the UFOs, they appeared to be proving their own theories wrong. Mantell, for example, may have died while chasing a skyhook, but it was certainly not possible to dismiss all unidentified flying objects as huge metallic-looking balloons. On one occasion near the top secret White Sands Proving Grounds in New Mexico, the crew of an aircraft was tracking a skyhook when they saw two disks fly in from just above the horizon, circle the balloon at nearly 90,000 feet, and rapidly depart. When the balloon was recovered it was ripped. Captain Ruppelt, who published this account, does not specify the date but it would have been in either 1948 or 1949. Compare it with the experience of six observers in 1951. Two of the witnesses were from General Mills, whose aeronautical division had launched and tracked every skyhook balloon prior to mid-1952. The other witnesses were from Artesia, New Mexico. The group had been watching the movements of a skyhook for about an hour when one of them noticed two specks on the horizon. He pointed them out because two aircraft were expected at Artesia airport at about that time. But the specks were not aircraft. They flew in fast in close formation, heading straight for the balloon. They circled the skyhook once and flew off in the direction from which they had come, disappearing rapidly over the horizon. As they circled the skyhook the two dull white objects tipped on edge, and the observers saw that they had a disk shape. Their estimated size was 60 feet in diameter.

People who believed that UFOs were natural phenomena, such as meteorites, were finding it difficult to convince others because eminent astronomers and scientists were among the UFO observers. For example, the man who reported a flight of six or eight greenish lights in Las Cruces, New Mexico in August 1949 was Professor Clyde W. Tombaugh, discoverer of the planet Pluto. New Mexico was being plagued by a special type of UFO known as green fireballs. Dr. Lincoln La Paz, world-famous authority on meteorites, and a team of intelligence officers made a full investigation of the sightings. Many of them saw the green fireballs—including La Paz—but no physical evidence was found to show that they were a special type of meteorite. In fact, Dr. La Paz concluded that whatever they were, they most certainly were not meteorites.

Physical evidence was what was missing from all the early UFO reports, however good. There were many excellent visual reports, but no photographs or film of sufficient quality to aid identification. There were intriguing radar reports of strange fast-moving objects. But without visual corroboration that an object was where

the radarscope recorded it, the report had to be dismissed as either a malfunction of the equipment or a misidentification of an echo caused by weather, birds, or other familiar objects. What the saucer buffs needed to prove their case was a sighting that combined visual and radar evidence. Gradually, such reports began to trickle through, building up to a flood, with an astonishing visitation in the skies to the Capital itself.

The spectacular began in the dying hours of July 19, 1952, when two radars picked up eight unidentified objects on their screens at Washington National Airport. Whatever the objects were, they were roaming the Washington area at speeds of between 100 and 130 miles per hour. They would suddenly accelerate to "fantastically high speeds," and leave the area. The long-range radar in Washington has a 100-mile radius, and was used for controlling all aircraft approaching the airport. The National Airport's control tower was equipped with a shorter range radar designed for handling planes in its immediate vicinity. Just east of the airport was Bolling Air Force Base, and ten miles further east was Andrews Air Force Base, which were also equipped with short-range radar. All these airfields were linked by an intercom system. All three radars picked up the same unknown targets. One object was logged at 7000 miles per hour as it streaked across the screens, and it was not long before the UFOs were over the White House and the Capitol, both prohibited flying areas. Radar experts were called in to check the equipment, though it was clear that the odds against three radarscopes developing identical faults were exceptionally high. They were found to be in good working order. Visual sightings were made as airline pilots came into the area and saw lights they could not identify. The lights were exactly where the radars were

Above: a drawing of the fireball seen by Dr La Paz and his wife. La Paz said that the fireballs were too big, their trajectory too flat, and their color too green for them to be ordinary meteors.
Below: Dr Clyde Tombaugh, the discoverer of the planet Pluto. He also saw the mystery lights.

Above: an artist's melodramatic impression of the "Washington Flap" of July 19, 1952. On that night airport radars picked up a formation of UFOs flying over the White House and the Capitol.

Below: a group of glowing objects in the sky over Salem, Massachusetts taken by a Coast Guard photographer on July 16, 1952. It was released by the Coast Guard at the height of the UFO craze.

picking up UFOs. A commercial airline pilot was talking to the control tower when he saw one of the lights. "There's one—off to the right—and there it goes," he said. As he reported, the controller had been watching the long-range radar. A UFO that had shown to the airliner's right disappeared from the scope at the very moment the pilot said it had.

One of the best ground sightings that night came when the long-range radar operator at the airport informed Andrews Air Force Base tower that a UFO was just south of them, directly over the Andrews radio station. When the tower operators looked out they saw a "huge fiery orange sphere" hovering in the sky at exactly that position.

Incredible as it may seem, no one thought to inform ATIC, and it was nearly daylight when a jet interceptor arrived over the area to investigate the phenomena. Its crew searched the skies, found nothing unusual, and left—but the UFOs had left the radar screens by then in any case.

A week later almost to the hour, the flying saucers were back over Washington to give a repeat performance. The same radar operators picked up several slow-moving targets at about 10:30 p.m. on July 26. The long-range radar operators began plotting them immediately. They alerted the control tower and Andrews Base, but it already had them on its screens and were plotting them. A call went out for jet interceptors. Once again there was a delay but two jets finally arrived soon after midnight. The UFOs mysteriously vanished from the screens just as the jets arrived. The pilots could see nothing during their search, and returned to base. Minutes after the jets left the Washington area the UFOs came back! The jets were called back, and this time when they reached the area, the UFOs remained. The controllers guided the pilots toward groups of targets, but each time the objects flew away at great speed before the pilots could see more than a strange light. At times it seemed that the lights were somehow monitoring the control tower conversations with the pilots, and responding before an aircraft had time to change course. On one occasion a light remained stationary as the jet pilot flew at it with full power. As he closed in the light suddenly disappeared, like a light being switched off. Both Washington sightings lasted several hours, and the second one gave some of the Air Force's top saucer experts an opportunity to rush to the airport, watch the radar screens, and listen to the pilots' accounts.

After growing pressure from the press and public the Air Force held a press conference—the largest and longest since World War II. It was presided over by Major General John Samford, Director of Intelligence at the Pentagon. He offered no concrete explanation, but hinted that the lights came from strong temperature inversions, during which light waves are refracted or bent. Therefore an object on the ground or the moon, sun, or stars may be reflected on a layer of inverted air. Similarly, temperature inversions could cause activity on radar screens.

However, the men controlling thousands of lives in and out of Washington National Airport had had years of experience, and were used to seeing and identifying every type of radar blip. They knew about temperature inversions, and were certain that they were not responsible for the radar UFOs whose blips came from

The "Washington Flap" of 1952

Left: Harry Barnes, the chief of radar at Washington's CAA control center where the mysterious blips were tracked, working over the radarscope that picked up the strange and unaccounted-for objects.

Below: an official Air Force diagram showing the position and course of the UFOs tracked over National Airport, Washington, D.C. during the night of July 26, 1952, a week after the first UFO sightings over the capital.

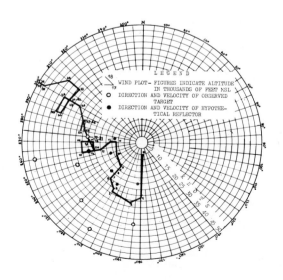

solid objects. Besides, as Captain Ruppelt admitted much later, the temperature inversions known to exist on the nights in question were not strong enough to affect radar in the way they normally do. In fact, mild inversions existed on practically every night of June, July, and August 1952. The sensational Washington sightings therefore remain unexplained.

The 1950s progressed without any significant lull in flying saucer sightings. More good sightings were added to the bulging files, but none proved sufficiently outstanding to convince officialdom that the saucers were interplanetary. The public was easier to convince, and the ranks of the believers in UFOs grew all the time. Many asked what it would take to convince the authorities. The answer seemed to be nothing short of a mass landing on the White House lawn.

In August 1953 the Air Defense Command radar station at Ellsworth Air Force Base near Rapid City, South Dakota, picked up a "well-defined, solid and bright" target at the spot where a ground observer had reported seeing a light. Height-finding radar also recorded it at 16,000 feet. The object was almost stationary. Then it picked up speed, circled Rapid City, and returned to its original place in the sky. This maneuver was watched on radar and reported by the ground observer.

Flying saucers were not just seen in the United States. UFO reports came from all around the world.
Above: a "saucer nest" found in Queensland, Australia. A farmer claimed he had seen a spaceship that rose up out of a swamp. When he investigated, he found this "nest" with the reeds all flattened in a clockwise direction.

Below: an artist's impression of a UFO that hurtled toward a DC-3 flying over Sweden. It finally shot underneath the plane.

A jet hurried to the scene, and soon spotted the light. It closed in to within three miles of the object. Then the light sped off with the jet on its tail. There was a limit to how close the UFO allowed the aircraft to come before pulling away, as though its power supply was automatically linked to a device measuring the distance of its pursuer. Captain Ruppelt reported what happened next in *The Report on Unidentified Flying Objects*:

"The chase continued on north—out of sight of the lights of Rapid City and the base—into some very black night.

"When the UFO and the F-84 got about 120 miles to the north, the pilot checked his fuel; he had to come back. And when I talked to him, he said he was damn glad that he was running out of fuel because being out over some mighty desolate country alone with a UFO can cause some worry.

"Both the UFO and the F-84 had gone off the scope, but in a few minutes the jet was back on, heading for home. Then 10 or 15 miles behind it was the UFO target also coming back."

The alert pilots of the jet interceptor squadron at the base had heard the conversation between the controller and pilot. Another jet with a veteran of World War II and Korea at the controls was standing at the ready. This pilot wanted to see a flying saucer for himself. Permission was given and he climbed into the sky, spotted the UFO, and closed in. Once again the object sped away, keeping a three-mile distance. The pilot switched on his radar-ranging gunsight which in seconds showed that there was something solid in front of him. At that point he got scared and broke off the interception. The UFO, on this

UFO Sightings Around the World

Left: a British saucer society staring hopefully up at the sky.

Left: a possible saucer landing site? The president of the Norwegian UFO center bends over one of the several mysterious radioactive triangular impressions found in the Nams Fjord in 1972.

occasion, did not follow him back to base. That encounter, according to Ruppelt, "is still the best UFO report in the Air Force files."

Washington National Airport was the scene of more UFO activity in July 1955 when a brilliant circular object with a tail four or five times its length flew into the airport, stopped and hovered. It oscillated before taking off at high speed, and was caught in the beam of a searchlight; but the searchlight suddenly went out. According to Brinsley le Poer Trench, a well-known British author of flying saucer books. "The UFO also caused the ceiling lights at the airport to go out, but these lit up again the moment the UFO left the airport."

When the first decade of flying saucer activity was over, the puzzle remained. There were photographs, radar reports, and thousands of sightings from experienced observers. But the experts were no nearer solving the flying saucer mystery than they were when Kenneth Arnold's sighting hit the world headlines. Then the Soviet Union launched the first man-made satellite on October 4, 1957.

Humankind was on the verge of exploring space—and sending our own·UFOs to other worlds.

The UFO Escort

The Stratoliner of the British Overseas
Airways Corporation (now British Airways)
was three hours out of New York on its run
to London. At that point, Captain James H.
Howard and his copilot noticed some
strange uninvited company three miles off
their left side: a large elongated object and
six smaller ones. These UFOs stayed
alongside for about 80 miles.

As the plane neared Goose Bay, Canada
for refueling, the large UFO seemed to
change shape, and the smaller ones
converged on it. Captain Howard told the
Goose Bay control tower what was
happening. Ground control in turn alerted
the USAF, which sent a Sabre fighter to the
scene. When Captain Howard contacted the
Sabre pilot, he said he was coming from
20 miles away.

"At that," said Captain Howard, "the
small objects seemed to enter the larger, and
then the big one shrank." He didn't find out
what finally occurred because he had to
leave Goose Bay on schedule.

These events took place in June 1954.
The pilot and copilot described the UFOs as
spaceships, and were confirmed in this belief
by various passengers. However, in a 1968
report by a USAF-sponsored research team,
this close-range sighting was dismissed as
an "optical mirage phenomenon."

Chapter 7
The Conspiracy of Silence

As flying saucer reports began occurring regularly, the public looked to the government to explain what was happening. The United States Air Force began a series of investigations in the attempt to discover whether any of the reports could be considered genuine, and if so, what possible explanation there could be. But did they set out on this investigation with an open mind, or was the conclusion decided in advance? The controversy rumbles on. Was there an official attempt to cover up the facts? How were the investigations conducted? Were there sightings of UFOs which even the investigating committees accepted as inexplicable?

Mass hysteria gripped the United States as thousands of citizens prepared for a Martian invasion. People fled madly from their homes and many of them reported seeing the invaders. This widespread panic, however, had nothing to do with flying saucers. It happened in 1938, nine years before flying saucers were reported, and it was caused by a radio dramatization by Orson Welles of H. G. Wells' science fiction story *The War of the Worlds*. The broadcast opened as though it were a newscast. Thousands of Americans, tuning in late, believed they were listening to a real news flash—and did not wait to hear more.

In the 1960s a similar reaction occurred in England when a TV drama included a mock news broadcast about a satellite that had been spotted in an orbit which kept it stationary over the British Isles. The make-believe news bulletin indicated that it might be armed with a nuclear weapon. People dashed out of their homes to warn others of the impending enemy attack before they realized they were watching a play rather than a news report.

If radio and television drama can cause such panic, what would be the reaction of the public to a government announcement that an alien race from another planet was watching us from the skies? UFO enthusiasts declare that it was fear of just such a panic that persuaded the United States government to hide the true facts about flying saucers from the American people. But was there a conscious decision for a cover-up

Opposite: an illustration from the original magazine serialization in 1897 of H. G. Wells' novel *The War of the Worlds*. It was a radio dramatization of this story of terrifying creatures from outer space invading Earth that caused panic in the United States in 1938. The artist has depicted the horrifying moment when the space creatures climb out of their ominous machines.

among top-level personnel, or were they just as confused and puzzled as everyone else?

Flying saucer reports began to occur with regularity about two years after the end of World War II. People were still tense and jittery in the West from the effects of the War, and the strained relations with Communist countries added to everyone's feelings of unease. It was a period of diplomatic and military uncertainty, and in the United States especially the Air Force was nervous and jumpy about reports of strange objects appearing in the sky. They had the responsibility of discovering whether the UFOs were simply wild misidentifications of natural phenomena, hostile aircraft on spying missions of whether the objects were, indeed, strange spacecraft manned by beings from outer space.

As the reports began to build up, the Air Force itself became alarmed by the possibility of aircraft of unfriendly states spying on the country's defense system—and by the possibility of the continent being kept under observation by unknown visitors from outer space. Quickly, the Air Force set up its own investigation into the reports of the unidentified flying objects and by as early as September 1947 the investigatory committee was meeting under the guidance of the Air Technical Intelligence Center (ATIC). Known by the various code names of Project Sign, Project Grudge, and Project Blue Book, the investigation lasted 22 years. It was finally wound up in December 1969. In the event,

Above: a 1946 U.S. Army test of a German V-2 rocket captured in World War II, carrying scientific instruments instead of explosives. It was known that the Soviets had also captured such rockets, and it was widely believed that they were testing them over U.S. territory. This gave rise to the fear that UFOs were the creation of the Cold War enemy in preparation for action against the United States.

Right: the Focke-Wulf Tiebflügel, an upright, vertical takeoff aircraft still in the design stages at the end of the war. At the end of each of the three long arms of this technologically advanced craft was a small jet propulsion unit. The rotating arms lifted the device from the ground like the blades of a helicopter.

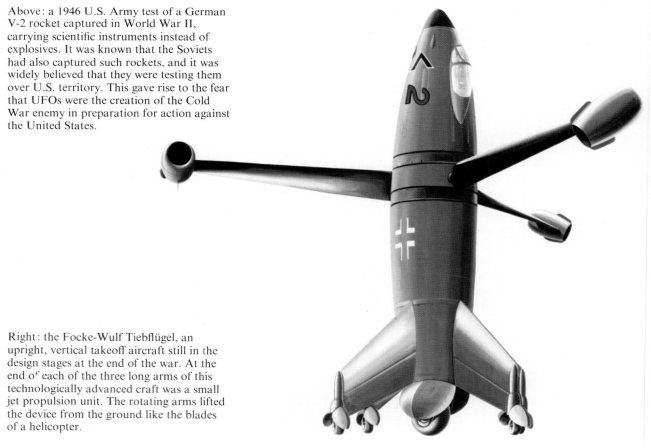

the United States was the only country to take the problem of UFOs seriously enough to set up an investigation of this kind. An immense amount of time and money were spent over those 22 years in trying to establish the facts. But, partly through bad public relations, partly through arbitrary changes in policy, the whole investigation could be said to have backfired.

In the early days almost any object not readily identifiable by ground observers stood a chance of making headlines as a flying saucer in the local papers, and the Air Force was faced with numerous reports. But when all the reports had been sifted through, and those due to misidentification of known objects had been eliminated, there still remained a sizeable number for which there was no clear explanation. It was then that Air Force policy was at its weakest. While privately pursuing its own serious investigations into these sightings, it publicly derided the possibility of the existence of unidentified flying objects and attempted to fob off the press with feeble and unsatisfactory explanations. These explanations were thought to be just as unconvincing and contradictory by the public, which was generally more sophisticated and cynical than had been realized. It seemed just too strange to many members of the public that such observers as trained airforce and civil aviation pilots should suddenly start mistaking flocks of birds or reflections from their own planes as flying saucers when they never had done so before.

UFO's – or Secret Weapons?

Other captured German equipment also fed fears of a Soviet attack.
Below: a German jet captured in the Allied advance in 1945. The GIs who found it abandoned, with bombs ready for loading, reported it "has no propeller and its nose is shaped like a shark's."

The Air Force Flying Saucers

In 1950, in the face of Air Force denials that flying saucers existed at all, 94 percent of people interviewed in a nationwide poll in the United States said they believed in unidentified flying objects. Whether they thought that the Air Force was hiding the truth or was just too stupid to recognize it is not entirely clear. But one thing was certain—there was a yawning credibility gap between what the authorities told the public and what the public believed.

Edward J. Ruppelt, leader of Project Blue Book for two years, tried to establish what actually happened in the first years of the flying saucers. In his book *The Report on Unidentified Flying Objects*, he maintained that the conflicting statements issued to the press and the air of confusion were the genuine results of bewilderment and lack of coordination among Air Force staff. He didn't think the military was deliberately hiding the facts behind a smokescreen, as some critics later believed, but he admitted that if the Air Force had "tried to throw up a screen of confusion, they couldn't have done a better job."

Before setting up Project Sign in September 1947, the Air Technical Intelligence Center had made a preliminary study of flying saucer sightings, and had come to the conclusion that unidentified flying objects were real phenomena. Project Sign

Below: the V-173, an experimental wingless aircraft known as the "flying pancake." This picture was released by the Navy in July 1947 during the first UFO excitement. The Navy said it was the only craft operated at that time which could answer the descriptions of the flying disks being sighted everywhere. But most of the sightings were in the far west, and this craft had never left Bridgeport, Connecticut.

Above: this is the official picture of the AVRO "disk" that was released after the photograph hit the newspapers. A kind of hovercraft, the AVRO was abandoned at the experimental stage.

Right: in 1959 Jack Judges, a freelance cameraman, snapped a picture as he flew over the A. V. Roe Aviation Company's plant, and his lens caught this saucerlike aircraft sitting on the ground. The mystery craft was publicized as a possible secret weapon—one which might account for many of the flying saucer sightings of the previous 12 years.

was accordingly given a priority rating and security coverage. Having accepted that flying saucers were real, the problem the top intelligence officers had to grapple with was their origin. After eliminating the possibility that they were a domestic secret weapon, possibly built by the Navy, there appeared to be two alternatives. Either they were some kind of spacecraft developed by the Soviet Union from Germany's leftover wartime designs, or they were interplanetary. All the German designs were carefully examined, but by the end of 1947 it was decided that there was no conceivable way in which the Soviet Union could have constructed a craft that behaved in the way unidentified flying objects did. The ATIC, still not questioning the assumption that UFOs were real, began to think that they must have come from outer space and that they had been built

The Start of the Investigations

Below: the Lubbock Lights as photographed by a student at the Texas College of Technology. The phenomenon was seen on 14 occasions in 1951 in one of the best authenticated sightings of UFOs. Negatives of five shots of the Lubbock Lights were tested intensively and, according to the official statement, were "never proved to be a hoax, but neither were they proved to be genuine." Edward J. Ruppelt, former leader of Project Blue Book and author of an exposé of the investigation, said that such well-documented sightings were not taken into account enough by the Air Fore.

Below right: General Hoyt S. Vandenberg, Air Force Chief of Staff in 1948. He was supposed to have received a top secret report from Project Sign, the Air Force's own investigation into UFOs, later known by the code names Project Grudge and Project Blue Book. According to rumor, the Project Sign report was destroyed because it concluded that UFOs were interplanetary vehicles, and General Vandenberg refused to accept this conclusion for lack of proof.

by a race with superior technological achievement. The problem that absorbed them was how to collect interplanetary intelligence. There were no appropriate guidelines or past experience for this task.

While these exciting ideas were being discussed in the Air Technical Intelligence Center, the Air Force continued to look foolish in its public stance. A statement put out from the Pentagon at about this time declared that flying saucers must be one of three things—reflections of the sun on low-hanging clouds, broken up meteors whose crystals had caught the rays of the sun, or large flattened hailstones that were gliding through the air. According to Ruppelt, a follow-up report was immediately issued which declared that these ideas were ridiculous. "No one had ever heard of crystallized meteors, or huge flat hailstones, and the solar reflection theory was absurd." No wonder the public was mystified and preferred to draw its own conclusions.

Why did the Air Force bother to issue ludicrous and unconvincing explanations that it was later obliged to retract? Was ATIC afraid of causing a panic if they voiced their real theories? Were they so unwilling to admit their own deep uncertainties that they took refuge behind an assortment of ill-considered excuses? Whatever the motives, the attitude of the Air Force with its claims and counterclaims, denials and false explanations, gave rise to a firm belief in a government conspiracy to cover up the truth about flying saucers. Arbitrary shifts in policy on flying saucers fanned this belief. After several years it had grown to such an extent that no serious attempt to study unidentified flying objects could be complete without an examination of the conspiracy claim.

It adds to the problem that most technologically advanced nations have secrets of some kind which they attempt to hide from other countries. Usually these secrets are known only to a few men. Even if the Air Force spokesmen who denied the existence of flying saucers believed what they said, the public

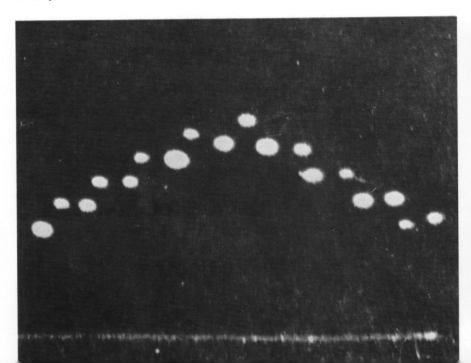

could not be sure that these spokesmen—or even their superiors
—had access to all the facts.

Many people at first believed the Air Force was pretending
not to know about the strange flying disks because they were in
fact a new top secret weapon. But as the years passed and UFO
reports came in from almost every country on the globe, it
became apparent that this reason could not be applied to all
governments.

In doubt that UFOs originated anywhere in the world, the
public began to speculate on whether they originated in Outer
Space. Air Force Intelligence meantime went through a series
of shifts in attitude. On July 24, 1948 an incident that shook the
staff of Project Sign took place. A large commercial plane
nearly collided with a UFO, and the pilots were able to give an
accurate description of it to ATIC investigators. No one at top
level appeared interested, however.

Despite the fact that UFO sightings and reports kept getting
better and better over the next months, it seemed that certain
people in authority would not accept their reality under any
circumstances. Gradually people working on Project Sign
found that the Pentagon did not want information that might
establish the existence of UFOs, but only explanations to prove
that they didn't exist. This attitude was confirmed in February
1949 when the code name of Project Sign was changed to
Project Grudge, and was staffed by those who would agree to
follow the new policy of explaining each sighting in terms of

Above: E. J. Sullivan speaking to a meeting
of the Civilian Saucer Investigations, one of
the earliest nonmilitary organizations
investigating UFO reports. Such groups
were formed because of official skepticism
about all flying saucers. Among CSI
members was Dr. Walther Riedel (behind
Sullivan), a renowned German rocket
designer.

Right: Dr. H. P. Robertson of the California Institute of Technology. He was the chairman of the CIA committee established in 1953 to investigate the UFO reports.

Far right: Donald E. Keyhoe speaking on UFOs in 1973. A retired Marine Corps general, Keyhoe had for many years maintained that UFOs are from outer space. His magazine article "Flying Saucers are Real" caused a sensation in 1949.

misidentification or hallucination. But the press releases and articles approved by Project Grudge debunking flying saucers seemed to have the reverse effect on a public which knew that a lot of extremely reliable people, including air line pilots and scientists, had seen unidentified flying objects. The general suspicion that the Air Force was hiding the facts increased, and by December 1949 the public was ready for the sensational article written by Donald Keyhoe and published by *True* magazine.

The article was called "Flying Saucers are Real." The author maintained that after eight months of research he had reached the conclusion that the earth was being watched by intelligent beings from another planet, who traveled in vehicles that we called flying saucers. He examined some of the most puzzling saucer cases and demolished the Air Force's official explanations. His authoritative style, and the fact that *True* had a reputation for printing the facts, gave his story great impact. UFOs zoomed back into the headlines, and the article was picked up and discussed by TV and radio commentators all over the country. Keyhoe had played the security angle very strongly in his story. He had talked to certain top level contacts who had denied that UFOs were any kind of secret weapon. From this he had deduced that they must be interplanetary spaceships, and that the Air Force was covering this up. According to Ruppelt, who later looked into the question, the Air Force was not covering up. They had simply lost interest in the question of UFOs, and couldn't believe that anyone else was still concerned. However, the idea of a government conspiracy of silence took root even more strongly among many members of the public, especially those who had themselves investigated UFOs.

At the end of December 1949, despite a number of recent good sightings and the new public interest generated by the Keyhoe article, the Air Force decided to reduce operations in Project Grudge. It released a final report containing 600 pages of analyses and appendices. Despite the fact that 23 percent of the sightings mentioned fell into the "unknown" category, the report concluded firmly that UFO reports all originated from a mild form of mass hysteria.

To add to the general confusion, *True* magazine in March 1950 published another story that caused a furore. This time it was by the man in charge of a team of Navy scientists at White Sands, New Mexico. In an article entitled "How Scientists Tracked Flying Saucers" the author, Commander R. B. McLaughlin, described several UFO sightings made either by him or his crew at White Sands. About one of these he wrote: "I am convinced that it was a flying saucer, and further, that these disks are spaceships from another planet, operated by animate, intelligent beings." The story was surprising in itself. Even more startling, though, it had received complete military clearance despite the fact that it contradicted every military press release that had been issued over the last two years.

Who was the public to believe? When people like Commander McLaughlin risked their reputation by declaring a belief in the interplanetary origin of unidentified flying objects based on personal experience, it carried far more weight than the non-sensical explanations offered by faceless spokesmen for the Air Force. Part of the confusion was probably due to the fact that even at the highest level one part of the Air Force seemed to have no idea what the other was doing about flying saucers.

"Flying Saucers are Real..."

Left: Major Hector Quintanella, one-time head of Project Blue Book, shown with some of the bogus items sent to him by believers in UFOs. Among them are ordinary radio parts and buckwheat pancakes. Quintanella was much criticized for his method of interrogation. Critics accused him of asking only the questions that would get answers he wanted to substantiate his theory, which was that all UFO sightings could be explained away.

Right: Dr. Edward U. Condon, a physicist and head of the panel of scientists commissioned by the U.S. Air Force to investigate the flood of UFO reports in 1966.

Far right: photograph of a UFO in flight over the Mount Palomar Observatory. It turned out on later investigation to be simply a defect in the film.

Below: are UFOs really clouds? Strange formations like this one over Mount Rainer, Washington, were one of the suggestions by the Condon Committee as a natural explanation for saucer reports. The original Arnold sighting, incidentally, was made in this very area near Mount Rainier.

This became apparent in the investigation that followed the Fort Monmouth incidents.

In the summer of 1951 many radar reports of unidentified flying objects came in, and interest in the UFOs began to pick up among some of the staff at the Air Technical Intelligence Center. But since radar was known to do strange things under certain weather conditions, the reports did not attract official interest. All this changed starting September 10, 1951. At 11:10 a.m. on that day, at the Army Signal Corps radar center at Fort Monmouth, New Jersey, a student operator was trying to demonstrate automatic tracking on a radar set to an important group of visitors. The set was able to track objects flying at the speed of a jet. He saw an object about 12,000 yards southeast of the radar station, but was unable to locate it with the automatic tracking device. It was going too fast for a jet, he claimed. The weather conditions were checked but nothing was found that might have caused a faulty reading. Twenty-five minutes later a jet pilot flying over Point Pleasant, New Jersey saw far below him a silver disklike object, about 30 to 50 feet in diameter. At 3:15 p.m. that afternoon the radar at Fort Monmouth picked up an object traveling at the immense height of 93,000 feet. It could also be seen as a silver speck in the distance. Two more UFOs were picked up by radar the following day.

A copy of the report of the sightings was sent to Washington, and the Director of Intelligence of the Air Force ordered an immediate investigation.

This was followed by a top level conference at the Pentagon where it turned out that many of the generals and colonels present had no idea that the active work on UFOs had been largely curtailed. Orders were given to revitalize Project Grudge, and Captain Edward J. Ruppelt was put in charge. The project once more began to gain official prestige, and by March 1952 had become a separate organization known as the Aerial Phenomena Group. The discouraging name of Grudge was abandoned in favor of the code Blue Book.

Investigations into UFOs seem to have been conscientious and unbiased under Ruppelt, but the Press remained suspicious of a cover-up. After the extraordinary UFO sightings at Washington National Airport in July 1952, the Air Force had to call a press conference. It had the unfortunate result of again sounding like a cover-up.

One of the most disturbing—though perhaps not altogether surprising—elements of the flying saucer controversy is the revelation that the Central Intelligence Agency was involved in setting policy for UFO investigation in the first stages. In the early 1950s the CIA undertook a study of UFOs, aided by a panel of five scientists under the chairmanship of Professor H. P. Robertson of the California Institute of Technology. The Robertson Panel's report was completed in 1953. It was partially declassified in 1966. The report concluded that UFOs did not present a threat to national security, but it observed that "the continued reporting of these phenomena does, in these parlous times, result in a threat to the orderly functioning of the protective organs of the body politic."

No Threat to National Security

Above: a drawing of a proposed U.S. Air Force project in which an aircraft would eject and light gases to produce a huge bubble of burning gas, as shown. The project was designed to test how people would react if they saw this kind of bubble, which behaved and looked like many UFOs. The point was to see if people not only reported the bubble as a UFO, but also reported space visitors in connection with it. The test was never carried out.

Below: the Barra Da Tijuca UFO in the first of the five photographs obtained. This reportedly shows the UFO as it approached.

What the panel feared was that channels of communication would become clogged with UFO reports. Even more serious, if an enemy could produce a UFO scare it might well be able to spring a surprise attack on the United States under cover of the flying saucers. The report went on to recommend a program that would "debunk" the UFOs, and educate the public in recognizing objects that might be mistaken for flying saucers.

There is no evidence that these recommendations were ever carried out. The Air Force continued to debunk saucers in public, but nothing was done to strip the subject of its aura of mystery. UFO reports continued to be filed by reliable and experienced witnesses and the number of sightings labeled "Unknown" grew to significant proportions.

What was needed was a completely impartial study of UFOs by independent scientists given access to Air Force files. UFO enthusiasts had demanded such an inquiry for years, and it came about as a result of a letter written by Major General E. B. LeBailly, then head of the Office of Information of the Secretary of the Air Force. It was dated September 28, 1965, and was addressed to the Military Director of the Air Force Scientific Advisory Board. After stating that 9265 UFO reports had been investigated by Project Blue Book between 1948 and mid-1965, and that 663 of these could not be explained, he added:

"To date, the Air Force has found no evidence that any of the UFO reports reflect a threat to our national security. However, many of the reports that cannot be explained have

Below: the Barra Da Tijuca UFO was sighted in May 1952 on the coast of Brazil by a reporter and a photographer. This sighting, one of the better known ones, was shown up as a fraud by the Condon Committee investigators. They noticed that the shadow on the disk could only have been produced if the sun had been shining directly *up* from the surface of the water.

come from intelligent and well qualified individuals whose integrity cannot be doubted. In addition the reports received officially by the Air Force include only a fraction of the spectacular reports which are publicized by many private UFO organizations.

"Accordingly, it is requested that a working scientific panel composed of both physical and social scientists be organized to review Project Blue Book—its resources, methods, and findings—and to advise the Air Force as to any improvements that should be made in the program to carry out the Air Force's assigned responsibility."

As a result an Ad Hoc Committee to Review Project Blue Book was set up. It recommended that the Air Force's UFO program should be strengthened "to provide opportunity for scientific investigation of selected sightings in more detail and depth than has been possible to date." After this, the University of Colorado agreed to undertake a study of UFOs under the direction of Dr. Edward U. Condon, a well-known physicist.

Here at last was the opportunity to find out what the UFOs were. The Colorado university study was to be an unbiased survey of the whole UFO scene conducted by experts in psychology, astronomy, radar, high-energy physics, meteorology, mathematics, astrophysics, aerophysics, chemistry, and many other relevant fields.

Most people who had distrusted the Air Force expected the truth from the Colorado project when it began its mammoth task of sifting the evidence on flying saucers in November 1966. But by the time its findings were released in January 1969, the belief that Dr. Condon and his fellow scientists were also party to a cover-up was widely held in flying saucer circles. Some civilian UFO organizations that initially enjoyed good relations with the Colorado project became dismayed by various events, and withdrew cooperation at an early stage. The official report seemed to justify their action. After more than two years of study the scientists concluded that what the Air Force had maintained since the earliest flying saucer sightings was true: flying saucers did not exist.

The Colorado project ran into credibility problems in the first months of its existence. Dr. Condon addressed a meeting at which, according to the *Star-Gazette* of Elmira, New York, he said: "It is my inclination right now to recommend that the government get out of this business. My attitude right now is that there's nothing to it . . . but I'm not supposed to reach a conclusion for another year. Maybe it [the UFO riddle] would be a worthwhile study for those groups interested in meteorological phenomena." That statement was made less than three months after the project began, and just a few months before Dr. Condon asked for about a quarter of a million dollars more to continue his work.

This statement upset many of the UFO societies. Then came another blow for those expecting an unbiased report. A memo written by Robert Low, the project administrator, was smuggled out. It was written when the UFO project was under consideration by the university, of which Low was Special Assistant

Establishment of the Condon Committee

Below: another of the Barra Da Tijuca pictures. In this one the shadow on the disk in the air is on the right side of the UFO, whereas the shadows of the trees and shrubs on the ground appear to lie on the left side.

Above: British policemen inspect one of six bleeping disks found in England in September 1967. A bomb disposal expert found a small loudspeaker, batteries, and several gallons of flour paste inside.
Below: two students, Chris Southall and David Harrison, admitted they had fabricated the "saucers" from the mold they are holding—as a fund-raising stunt.

to the Vice-President and Dean of Faculties. He wrote: "The trick would be, I think, to describe the project so that, to the public, it would appear a totally objective study, but to the scientific community, would present the image of a group of nonbelievers trying their best to be objective but having an almost zero expectation of finding a saucer."

It seemed that the report's conclusions were being shaped even before the university received the contract. Dr. J. Allen Hynek, the astronomer who worked closely with Project Blue Book for many years, has this to say about the early days of the Colorado project: "I remember my own dismay when, on the occasion of my visit to the committee, when the project was scarcely two weeks old, Low outlined on the blackboard for us the form the report would take, what the probable chapter headings would be, how much space should be devoted to each chapter, with an implied attitude that he had decided already what the substance and tone of the report would be."

Dr. Condon has dismissed the significance of the Low memo, explaining that he did not see it until 18 months after it was written, and that he disagreed with some points in it including the suggestion that the study should concentrate on psychological rather than physical aspects of flying saucer sightings.

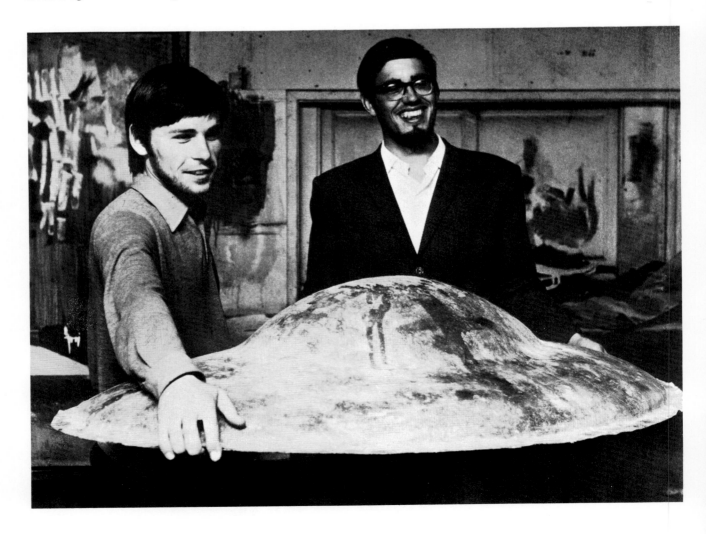

Nevertheless, since other commitments often kept Dr. Condon away from the project, Low was the man who, to quote Dr. Hynek again, "became the actual pilot of the Colorado UFO ship." Dr. Condon was sufficiently incensed by the disclosure of the Low memo, however, to dismiss the two committee members he regarded as responsible for the leak. They were pro-UFO, and without them the Colorado "UFO ship" could look forward to a smooth voyage with its nonbelieving passengers.

The final report of the Colorado UFO project is a massive tome—965 pages in its paperback version—which makes daunting reading for all but the most enthusiastic scientists and UFO followers. Seemingly in a helpful way, the report has its conclusions at the beginning so that readers not prepared to plow through acres of type can learn them in just over 40 pages. However, readers who give up at that point will be sadly misled, for Dr. Condon's summary avoids mentioning any of the outstanding cases that remained unsolved after the in-depth study by the Colorado scientists. Closer scrutiny of the entire report shows that the findings are based on a study of only 90 sightings out of a potential 25,000 available to them. What is more, these 90 cases are not the best or most puzzling.

Fake Saucers

Below left: is it a hovering UFO? Below: no, simply a metal plate. The photographer spins the plate into the air, snaps his picture, and he has his "flying saucer." Among the UFO pictures submitted to investigators have been many hoaxes as elementary as this one—a fact which skeptics are quick to point to as unfavorable evidence of the kind of person who is likely to make a UFO report.

Below: Paul Trent, a farmer of McMinnville, Oregon. He and his wife sighted a UFO in May 1950. His wife saw it first while feeding the rabbits at the back of the house, and shouted to him. He grabbed a camera in time to be able to get two pictures.

Above: Trent's first picture. The analysis of his negatives showed they had not been tampered with. Elaborate geometric and light calculations showed both pictures were consistent with each other and the time of day. Even the Condon report, having considered the possibility of a model hung from the wires, came to the conclusion that "this is one of the few UFO reports in which all factors investigated appear to be consistent with the assertion that an extraordinary flying object . . . flew within sight of two witnesses."

They are a curious cross section, including an attempt to rendezvous with a flying saucer which, it was said, would be landing at a specific time and place. It did not turn up! Other cases included lights and objects that any experienced investigator could explain away in minutes. In fact, 14 of the cases had already been evaluated as misperceptions by Project Blue Book.

Dr. Hynek, who is not a believer in the extraterrestrial origin of flying saucers, makes this observation about the Colorado findings: "The report opened with a singularly slanted summary by Dr. Condon, which adroitly avoided mentioning that there was embodied within the bowels of the report a remaining mystery; the committee had been unable to furnish adequate explanations for more than a quarter of the cases examined."

The press greeted the report as the UFO enthusiasts had expected. Basing their headlines on the summary, they announced: "Flying Saucers Do Not Exist—Official." The serious saucer investigators read between the lines and found what *they* expected: confirmation that UFOs exist. Many also regarded the discrepancies and method of presentation of the report as further evidence that the truth about UFOs is being hidden from the public.

The Colorado investigation did nothing to solve the UFO dilemma in general terms. From the Air Force's point of view, however, it was a useful exercise. Largely upon the recommendation of the Colorado team, the USAF terminated its official investigation of flying saucers in December 1969 with the closure of Project Blue Book.

Since then the public has been faced with a new dilemma. To whom should they report a flying saucer if they see one? Soon after the closure of Project Blue Book a letter went out from the Pentagon specifying: "The Aerospace Defense Command (ADC) is charged with the responsibility for aerospace defense of the United States. . . . Consequently, ADC is responsible for unknown aerial phenomena reported in any manner, and the provisions of Joint Army–Navy–Air Force publication (JANAP–146) provide for the processing of reports received from nonmilitary sources."

So, though flying saucers do not exist in the eyes of the Air Force, it would still like to know when people see them. However, under JANAP–146 heavy penalties are imposed on anyone disclosing information received about unidentified objects. So UFOs are still enveloped in a cloak of secrecy—or a conspiracy of silence.

Is the Air Force Hiding the Facts?

Below: Condon made little attempt to disguise his skepticism about the reality of UFOs—which did nothing to allay the belief in a conspiracy of silence—and was an easy target for satire. This 1967 Oliphant cartoon from the *Denver Post* plays on the director's attitude.

'STAY CALM, DR. CONDON—JUST TELL THEM YOU DON'T BELIEVE IN THEM!'

Chapter 8
Chariots of the Gods

Are many of the enigmatical constructions that puzzle archaeologists of today really the traces left by prehistoric astronauts who visited this Earth? The Swiss author Erich von Däniken has argued that this is the case in a series of books, drawing evidence from all over the world. His work has been widely criticized by more orthodox scholars, but the questions he raises are fascinating and challenging. Are there landing strips for spacecraft in the remote plateaus of the Andes? Do Australian aboriginal rock paintings depict ancient space travelers? Is there a prehistoric stone relief in Mexico that shows a spaceman in his rocket?

In the Peruvian Andes in South America lies the arid plain of Nazca, a strip about 37 miles long and a mile wide. The extraordinary feature of the land is that it is covered with strange geometric patterns formed by deep straight furrows that expose the yellowish subsoil. Some of the lines run parallel for miles. Others intersect or join to make large trapezoidal shapes. Among the lines numerous animal figures have been etched, the biggest being 275 yards at its widest point. But the odd thing about the figures and the geometric patterns is that they can only be properly appreciated from a height—for example, by someone flying over in a plane. Yet the people who planned the patterns and animal figures centuries and centuries ago were not able to fly. Why, then, are the designs so obscure at ground level? What were they for? What do they mean?

One archaeological theory is that the lines are Inca roads. But why should the Incas need such a complex system of parallel and intersecting roads, all of which terminate at the edge of the plain? Another supposition is that they are irrigation channels. But what purpose do the vast animal figures serve between the channels in that case? Perhaps more convincing is the belief that the lines represent some astronomical plan that may in some way be linked with a calendar. By far the boldest idea, however, and the one that has attracted most public speculation, comes from the Swiss author Erich von Däniken. He believes that the markings were designed to be seen from the air because the plain

Opposite: a strange ceramic figure found in South America. This is the kind of evidence that Swiss author Erich von Däniken uses to argue that space people visited the Earth in antiquity.

Amazing Theories of Von Däniken

Right: Erich von Däniken, author of sensational best-selling books about space people as the possible ancestors of humans, shown lecturing in the United States.

Below: an artifact of gold from Colombia, similar to many gold objects that have been found in other places in South America. Some theorists have claimed that it is a model of a delta-winged aircraft with cockpit, windshield, a flanged tail, and elevators.

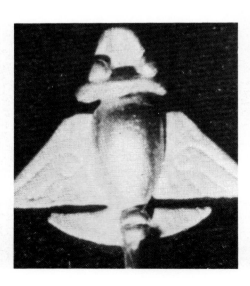

of Nazca was intended to be an airport. His theory is that many thousands of years ago inhabitants of another planet landed their spacecraft on the Nazca plain and improvised an airfield there.

Von Däniken has become world-famous on the basis of such bold suggestions. A self-taught man, he offers rather startling theories to explain the numerous mysteries that perplex minds molded by more orthodox teachings. He believes that all over the world evidence can be found to show that beings from outer space have walked on earth and talked to humans.

Nazca is just one piece of his jigsaw puzzle. He has filled four books with the other pieces, and his globetrotting research continues in an attempt to provide conclusive proof. These books, *Chariots of the Gods?*, *Return to the Stars*, *Gold of the Gods*, and *In Search of Ancient Gods* had worldwide sales approaching 30 million in 1975. The remarkable appeal of his theories has

Above: one of several sites near Nazca, Peru said to be landing fields for spacecraft by von Däniken. Since they can only be plotted from the air, he believes they were built with guidance from an aircraft. One of his critics remarked that an aircraft needing the length of some of the runways —which extend for miles—would have had impossibly bad brakes.

Airfield for the Gods?

In the 1930s some mysterious marking were discovered on a plain near Nazca, Peru. They consisted of long straight lines in a geometric pattern, with enormous animal figures between them. One of the most amazing things about these markings was that they had to be seen from the air to get a full overall picture. Yet archaeologists agreed that they were made long before aircraft existed. Were the lines ancient roads or irrigation canals as most scientists suggested? Or was the layout an airfield built by space people way in the past? This is the view of Erich von Däniken, whose writings on spacecraft landings in antiquity have been a sensational success.

Von Däniken believes that beings of superior intelligence landed near Nazca from space and built an airfield for future use. These "gods" were admired by tribes who saw them, and who hoped for their return. When the gods didn't come back, the tribes began to extend the airfield to entice them. First they just added lines. Then another generation thought of adding more lines according to the stars. Still later generations did the drawings as special landing signals.

That, von Däniken claims, is how Nazca became an airfield for the gods.

"Memory of Space Visitors"

Opposite top: a drawing of the Palenque relief. Von Däniken says that the pointed vehicle ending in a tail is clearly a rocket, and points out that the figure handling it has the right clothing for space travel.

Opposite bottom: a photograph of the stone relief. The sculpture is in a spot so narrow that there is no room to get a picture head-on. All photographs of it therefore have a distorted appearance.

Below: Temple of the Inscriptions in Palenque, Mexico. It has the famous stone relief that von Däniken says was of an astronaut.

been diagnosed by skeptics as "Dänikenitis." But strong criticism of many of his statements by authorities has done little so far to diminish the popularity of his beliefs.

Von Däniken differs from so many classical scholars in his insistence that the myths and legends of our distant past should be taken more literally. He believes that our own tentative steps in space exploration have added a new dimension to our past so that we can now examine the stories of our earliest recorded history in the light of our new experiences. For example, why is it that the myths of most cultures refer to gods coming from the skies? Until von Däniken put forward his theories, this idea was regarded as symbolic. But according to the Swiss author, if another race of beings had landed on Earth in the past, the description of their visit would tally exactly with what we find in these myths. He believes that latent memories of those visits have driven us to mount massive space programs to reestablish contact.

The controversial writer offers an additional and even more sensational reason for our deep-seated drive to reach Outer Space. The space visitors, he says, are our forefathers. *Homo sapiens* was created by the "gods" by an artificial mutation providing a cross breed between themselves and the more humanlike apes who populated the earth at that time.

Von Däniken's hypothesis is based on evidence like the ancient stone relief discovered in Palenque, Mexico in 1935. This is described by him in *Chariots of the Gods?* in these words:

"There sits a human being, with the upper part of his body bent forward like a racing motor-cyclist; today any child would identify his vehicle as a rocket. It is pointed at the front, then changes to strangely grooved indentations like inlet ports, widens out and terminates at the tail in a darting flame. The crouching being himself is manipulating a number of undefinable controls and has the heel of his left foot on a kind of pedal. His clothing is appropriate: short trousers with a broad belt, a jacket with a modern Japanese opening at the neck and closely fitting bands at arms and legs . . . And there [is the headgear] with the usual indentations and tubes, and something like antenna on top. Our space traveler—he is clearly depicted as one—is not only bent forward tensely, he is also looking intently at an apparatus hanging in front of his face."

The author sees other relics of the past in the same light. Almost every ancient figure depicted with strange headgear is described by him as a spaceman equipped with antenna. To argue that it is simply a ritual mask or headdress—as many specialists do—does not dissuade von Däniken. He asks why, if this is so, the masks or headdresses were shaped in that way in the first place. He suggests that it is a ritual tradition specifically designed to keep alive our memory of space visitors.

To support his case von Däniken cites rock drawings all over the world. In the Tassili mountains in the Algerian Sahara, for example, there are hundreds of figures that according to von Däniken are wearing round helmets with antenna on their heads, and seem to be floating weightlessly in space. One particular drawing has a more specific interpretation for him. It is of a sphere with a series of four concentric circles. At the top of the sphere there seems to be a hatch with an aerial protruding, and from the side two hands are stretched out. Accompanying the sphere are five floating figures with tight fitting helmets. Von Däniken concludes that these paintings represent the memories of a visit to earth by beings from outer space, handed down for many generations.

If our planet has been the temporary home of superior beings from another planet, we would expect to find more substantial proof than rock paintings. The Nazca site has been offered as evidence, and so has the mysterious plateau of El Enladrillado in Chile. This was rediscovered in 1968, and has provided archaeologists with yet another puzzle. The two-mile-long plateau can only be reached by a three-hour journey on horseback. When visitors arrive they are greeted by an astonishing sight. The ground is covered with more than 200 giant rectangular stone blocks which are extremely smooth of surface. They range from 12 to 16 feet high and 20 to 30 feet long. Exactly 233 of them have been grouped to form what resembles an amphitheater at a glance. Who could have hewn and shaped such enormous blocks with such precision, and how could they have managed to bring them to such an inaccessible spot? Why were they there, and what were they for? Archaeologists have not yet come up with a convincing explanation. Von Däniken says that a strip of ground of about a thousand yards long among the huge stones was a landing strip. According to the Chilean newspaper *El Mercurio*, the leader of the scientific expedition to investigate the plateau in 1968, Humberto Sarnataro Bounaud, said he believed that the stones were the work of an ancient unknown culture of advanced technology, because the natives of the region were incapable of such achievements. He also observed that whoever was responsible knew that the plateau would make a first-class landing ground for all kinds of flying craft. If this were so, the 233 symmetrically arranged stone blocks may have been visual landing sights for aircraft rather than an amphitheater.

Could superior beings with flying machines have shaped the blocks and positioned them, either to serve as a sign to aircraft or for some other reason? Our planet is scattered with strange and enormous buildings and monoliths. We find it hard to understand how they were erected or why they were positioned in certain sites. On tiny Easter Island 2250 miles from the coast of Chile are hundreds of gigantic statues, some between 33 and 66 feet high. They are a great puzzle. Who carved them? Why were they erected on this remote and inaccessible place? Questions also remain about how the Egyptian and Mayan pyramids were constructed. Nor do we yet know the meaning of giant stone circles like Stonehenge in the south of England.

If, as von Däniken suggests, a race of extraterrestrials was responsible for building many of these monuments then the mystery is to some extent solved. They would choose sites that marked places of special importance to them or served as landing strips.

Tiahuanaco is another South American wonder to which von Däniken refers. It is the site of a huge ancient city on a plateau at a height of 13,000 feet. The oxygen level is very low and working conditions must have been extremely hard. Despite

Worldwide Evidence

Below: a rock painting in the Tassili mountains, which lie in the Algerian Sahara. According to von Däniken the main figure, which seems to float weightlessly in empty space, was a Martian on a visit to Earth long years ago.

this, enormous buildings of some kind were constructed. The walls consist of huge blocks of sandstone weighing a hundred tons according to von Däniken. These are topped with other blocks weighing 60 tons. Large squared stones with smooth surfaces are neatly fitted together and fastened with copper clamps. One of the greatest archaeological wonders of South America is also found in Tiahuanaco. It is the Gate of the Sun. This gigantic 10-ton sculpture, carved out of a single block, tells the story of a golden spaceship that came from the stars. It carried Oryana, the woman who became the Great Mother of the Earth. She had four webbed fingers on each hand. After giving birth to 70 children, she returned to the stars.

Also found at Tiahuanaco was a calendar, which von Däniken describes as giving the position of the moon for every hour, taking the rotation of the earth into account. Still another extraordinary discovery was the Great Idol, a block of red sandstone over 24 feet long. It is inscribed with hundreds of symbols which, according to the book *The Great Idol of Tiahuan-*

aco by H. S. Bellamy and P. Allan, records an enormous body of astronomical knowledge based on a round Earth. Where did such knowledge come from?

To support his theories von Däniken cites a fascinating discovery in China. In 1938 the Chinese archaeologist Chi Pu Tei discovered a series of graves arranged in rows in the mountain caves of the Sino-Tibetan border district. The cave walls were decorated with figures in round helmets and the sun, moon, and stars linked together by groups of small dots. Chi Pu Tei and his assistants managed to salvage from the graves 716 granite plates, about two centimeters thick. According to von Däniken, these looked much like our long-playing records. The stone plates had a hole in the center from which a double-grooved incised script spiraled outward to the edge.

The disturbing thing about the graves was that, whereas the skeletons were small in stature, the skulls were large and broad. Chi Pu Tei suggested that these were the remains of a now extinct species of mountain ape—though apes are not known for making graves, particularly in neat rows. Clearly they were not responsible for the stone plates. The archaeologist suggested that the plates must have been put there at a later date by others of a different culture.

For 20 years experts tried to solve the riddle of the stone plates. Then in 1962 Professor Tsum Um Nui of the Academy of Prehistoric Research in Peking claimed he was able to decipher parts of the scripts. The Academy had doubts about the scripts and the astonishing story they told, so it refused to publish the findings. In 1963, when Tsum Um Nui had support from four other scientists, he published the results himself. The plates, he revealed, told of an aircraft that crashed on the third planet [Earth] 12,000 years ago. It did not have enough power to leave this world again and the people on board did not have the means to build a new aircraft. They tried to make friends with the mountain inhabitants but when some of them were hunted down and killed, the remainder hid themselves in the caves.

If space visitors wanted to explore our earth without being molested it seems likely that they would take shelter in natural caves, or even build their own subterranean world. It was this belief that took von Däniken to Ecuador to meet the discoverer of a huge complex of underground passages and halls beneath the South American continent.

Juan Moricz, an Argentinian, claims to be the legal owner of part of this strange underground world. But because he is convinced of the incalculable cultural value of his find he asked the State of Ecuador to take control and set up a scientific commission to study the area. No official response was forthcoming after Moricz deposited his legal title-deed, and the discovery awaits detailed investigation. But Moricz agreed to take von Däniken into the passages and show him some of their treasures. This experience is described in *The Gold of the Gods*.

The passages are not in a rough natural state. The walls and ceilings are smooth, and all form perfect right angles. One wonders what tools can have been used to carve these tunnels which, Moricz claims, run for thousands of miles beneath the earth's surface. The discoverer led von Däniken along the

Above: a figure of the type known as Wondjina to the Australian Aborigines who treasure them. They appear in rock paintings. The hieroglyphics are the same as those George Adamski claims he received from a Venusian in 1952.
Below: an ancient Mexican stone figure which von Däniken believes bears equipment of the modern astronaut. He uses this to support his contention that ancient civilizations were in contact with intelligent beings of considerable technological achievements.

Around the world are gigantic old
monuments that modern scholars have only
been partly able to explain why and how
they were built.
Above: the pyramid of Cheops. Why did
the Paraoh build it there?

passages and into a vast hall "as big as the hangar of a jumbo jet,"
with galleries branching off it. They continued through another
passage and into a perfectly proportioned but gigantic hall. In
the center was a table and chairs. They appeared to be made of
some kind of plastic but were as strong and as heavy as steel.

"There were animals behind the chairs: saurians, elephants,
lions, crocodiles, jaguars, camels, bears, monkeys, bison, and
wolves, with snails and crabs crawling about between them.
Apparently they had been cast in moulds and there was no logical
sequence about their arrangement . . . The whole thing was like
a fantastic zoo and what is more all the animals were made of
solid gold."

The most astonishing treasure was housed in the same animal
hall. It was a library of metal plaques and metal leaves only
millimeters thick. Von Däniken, who examined them, says that

most of them measured about 3 feet 2 inches by 1 foot 7 inches. The metal was unusual, he says, for the leaves stood upright without buckling, in spite of their size and thinness. They stood next to each other like bound pages of giant folios, and each was covered with writing that looked as if it had been stamped on by a machine. He estimated that there were between two and three thousand of these metal pages in an unknown writing. Von Däniken declares: "This metal library was created to outlast the ages, to remain legible for eternity." He believes the writing to be the oldest in existence on this planet, perhaps containing a history of man from his earliest days—and the visitation of extraterrestrial beings who left this permanent record of their intervention in the development of life on Earth.

In *Chariots of the Gods?* von Däniken offers this idea of what may have happened when an unknown spaceship discovered our planet:

"The crew of the spaceship soon found out that the Earth had all the prerequisites for intelligent life to develop. Obviously the 'man' of those times was not *homo sapiens*, but something rather different. The spacemen artificially fertilized some female members of this species, put them into a deep sleep, so ancient legends say, and departed. Thousands of years later the space travelers returned and found scattered specimens of the genus *homo sapiens*. They repeated their breeding experiment several times until finally they produced a creature intelligent enough to have the rules of society imparted to it."

He finds support for this theory in the story of Adam and Eve. Man existed, but woman had to be created. Is this a symbolic way of explaining that men came to the Earth and created a species? Why did they use Adam's rib? Von Däniken asks why foreign intelligences with a highly developed science who knew about the immune biological reactions of bones could not have used Adam's marrow as a cell culture and brought male sperm to development in it.

As von Däniken's books have flourished, so have the theories of this prolific Swiss author changed to fit new discoveries. He has tried to make every modern mystery fit into his extraterrestrial jigsaw. He flies around the world to study archaeological and

Ancient Monuments or Space Beacons?

Above: the ambiguous stone monoliths of Stonehenge in rural England.

Left: the Easter Island statues. How and why were they erected?

The Riddle of Easter Island

The colossal stone statues of Easter Island in the South Pacific have astounded and puzzled explorers, anthropologists, and ordinary tourists for hundreds of years. No one can explain how such huge stone slabs could have been carried from the only quarry on the island some distance away, nor how the volcanic stone could have been worked with the small primitive tools that are the only ones ever found on the island. Even more mysterious is the question of why the statues have long straight noses, narrow lips, and low foreheads when no island dwellers—nor any of their ancestors —look like that.

Erich von Däniken has a theory that fits in with his ideas about unexplained statuary, drawings, and monuments all around the world. The Easter Island figures, he says, were the work of space visitors.

According to von Däniken the space people were stranded on the island through some mishap. While waiting to be rescued— and perhaps as a signal to their rescuers— they built some gigantic statues and lined them up where they could be seen from afar. They could do all this easily because of advanced technology.

The gods left Easter Island suddenly, von Däniken suggests, which is why some of the statues remained unfinished.

Mysterious Monoliths

Below: the Gate of the Sun, part of the remains of the ancient city of Tiahuanaco in Bolivia. Said to have been carved out of a single block of stone, it weighs 10 tons. Forty-eight square figures in three rows surround a flying god. According to von Däniken, legend says that the city was created by a woman from a golden spaceship, who came from the stars to become the Great Mother of the earth. She gave birth to 70 earth children and then returned to her starry home. For von Däniken this legend supports his argument that giant projects like the Gate of the Sun were constructed with the aid of inter-planetary visitors.

historical wonders for himself. By the time *The Gold of the Gods* appeared his jigsaw was virtually complete. He believes that the prehistory of humankind happened something like the following:

In the distant past a battle occurred in our galaxy between humanlike intelligences. The losers escaped in a spaceship. To deceive their opponents they landed on Earth, a far from ideal planet, where they had to wear helmets and breathing apparatus until they accustomed themselves to the air. They then set up technical stations and transmitters on a different planet, then known as the fifth planet, and sent out coded reports. Their trick was successful.

The victors of the galactic battle were fooled into believing that those they had defeated had established a base on the fifth planet, and annihilated it in a gigantic explosion. This accounts for the existence of a planetoid belt containing thousands of lumps of stone between the present day fourth and fifth planets, Mars and Jupiter.

Meanwhile the space visitors burrowed deep into the ground on Earth and made tunnel systems that not only hid them from their pursuers, but also protected them from a general air attack. The victors, believing they had destroyed their enemies, withdrew to their home planet. But the imbalance in our solar system caused by the destruction of a planet made the Earth's axis move a few degrees out of position, and brought about the floods and deluges which are part of the legends of people all over the world.

When the space visitors eventually emerged from their magnificently built underground world they began creating intelligence on earth, using their knowledge of molecular biology. Taking the most advanced species of ape then in existence, they used it to create man in their own image. They laid down strict laws to ensure that the new breed remained pure, and they wiped out those who did not live up to their expectations. They were looked upon as gods by the Earth creatures. After generations of breeding, man as we know him emerged on our planet. That, in essence, is von Däniken's strange hypothesis.

No one would deny the right of revolutionaries and visionaries to hypothesize to their heart's content, even on matters of such fundamental importance as human origin. But scholars are always wary of the way supporting evidence can be bent to fit even the wildest theory, and they have accused von Däniken of such behavior. It is said that he has plundered almost every culture to find proof, and that he has disregarded all the facts that tend to disprove his claims. After a group of academics studied his first book *Chariots of the Gods?*, their often amusing comments were edited by E. W. Castle and B. B. Thiering and published under the title *Some Trust in Chariots!* They point out numerous fallacies on which von Däniken's theories are based, and on the basis of the most modern findings in the realms of history, archaeology, engineering, and religion, offer more mundane explanations for many of the feats accredited by von Däniken to the hypothetical space visitors.

In their introduction the editors put the case against von Däniken in these words: "Some of his 'facts' are false and his arguments from these 'facts' are not acceptable arguments at all." It seems that von Däniken can be attacked at many levels. Gordon Whittaker, an expert on the Aztec civilization, points out a mistake that could have been avoided by a quick perusal of some museum catalog. Von Däniken asserts that the giant heads sculpted by the Olmec people of Mexico are too heavy to ever be moved from their site because no bridge could stand their weight. But Whittaker shows that, not only do several museums possess such heads in their collections, but also one of the heads was recently transported thousands of miles for an exhibition at the Metropolitan Museum of Art in New York. At a different level, Whittaker also demonstrates that von Däniken's belief that the Easter Island inhabitants could not have carved and erected their giant statues unaided is ill founded. The statues are not the only huge monuments found on Polynesian islands, and there are native accounts of how they were moved with ropes and stone rollers.

Above: the Great Idol monolith at Tiahuanaco. The symbols which decorate the surface have been interpreted as recording an incredibly detailed knowledge of astronomy—astonishing in view of what could be expected of that period and in that isolated place.

Scholars Object

Below: a reconstruction of how a step pyramid may have been built, according to one theory of orthodox archaeology. Although von Däniken seems to think that rope did not exist, and that there was no wood in Egypt to use for rollers, there are in fact examples of Egyptian rope in existence. Moreover, records of cargoes of imported wood are well known to scholars. Von Däniken also appears to be mistaken about the weight of the blocks. He estimates the weight of each block as being 12 tons, but most of the blocks really weighed only about 2½ tons each.

Another contributor to this fascinating book that tries to take von Däniken's theories apart piece by piece is Professor Basil Hennessy from the University of Sydney, Australia. He refers to von Däniken's "fascinating collection of unsupported, undigested, disconnected and often inaccurate claims," and then goes on to show how seriously the Swiss author errs in his facts. For example, the Great Pyramid of Giza is said by von Däniken to weigh 31,200,000 tons. Professor Hennessy points out that there are an estimated 2,300,000 blocks known to weigh an average of 2½ tons for a total of 5,750,000 tons. Moreover the blocks are made of soft limestone rather than hard granite as von Däniken suggests, and the stone was quarried in the immediate vicinity of the pyramids and not miles away. The blocks were moved into position by rope, which von Däniken believes was nonexistent at that time but which can be seen in abundance in a number of museums. Von Däniken says that there was no wood to make rollers for moving the massive blocks, but Professor Hennessy reveals that the ancient Egyptians imported vast quantities of wood from other countries.

Dr. A. D. Crown of the Department of Semitic Studies at the University of Sydney challenges von Däniken on his theories about the Piri Re'is map, said to have been drawn by Admiral Piri at the beginning of the 16th century. Quoting American cartographers, von Däniken stated that when the map was constructed on a grid and transferred to a modern globe it was absolutely accurate. It showed the Mediterranean and Dead Seas, the coasts of North and South America, and even the contours of the Antarctic. The Piri Re'is map is said to have contained details of interiors as well as coasts, including the mountain ranges of the Antarctic which were not discovered until 1952 by echosounding apparatus. Von Däniken declares that aerial photographs must have been used as the basis for the Piri Re'is map. He suggests that these photographs were taken centuries ago from an aircraft or spaceship hovering high above Cairo. Copies were made and handed down, and copies of these copies were made, and so on. According to von Däniken the Piri Re'is map is a copy of a copy of a copy, but it is originally based on aerial photography.

Dr. Crown agrees that the map is impressive, but he shows conclusively that its origin is not a mystery. Nor is it necessary to believe in aerial photography to explain its existence. It does not show as much as von Däniken believes it shows, and what it does show was known to experienced mariners of the early 16th century, he says.

In a review of von Däniken's book *In Search of Ancient Gods*, Jacquetta Hawkes in the English *Sunday Times* of September 15, 1974, points out an error about the position of our galaxy that occurs on the very first page of the book. A second error concerns the dating of Tiahuanaco as 600 B.C. by von Däniken, whereas it should be A.D. 600. These, she wrote, are just "two out of the scores of simple errors scattered on almost every page." She also points out that more than 40 of the photographs and illustrations in the book are of well-known forgeries.

Experts are not always right of course. Besides there are also some experts who support von Däniken's beliefs. However, few

Above: a museum exhibit of South American carvings. The one on the right is an Olmec head of the kind that von Däniken says, "they can only be admired on the sites where they were found, for . . . no bridge in the country could stand the weight of the colossi." Challengers of von Däniken ask him how, in that case, the carvings were brought to the museum.

Left: an Easter Island statue in the British Museum. Von Däniken points out that it would have been impossible for Easter Island's small population to have carved and erected hundreds of these strange figures. However, a historian argues that the figures are typical of religious carvings on other Polynesian islands. He also argues with the point that the job of carving and erecting was not done overnight, but over a long period of time, using tools that are still found in quantity on the island. Inhabitants could describe exactly how the giant figures were put into position to anthropologist Thor Heyerdahl when he carried out his investigations into the matter.

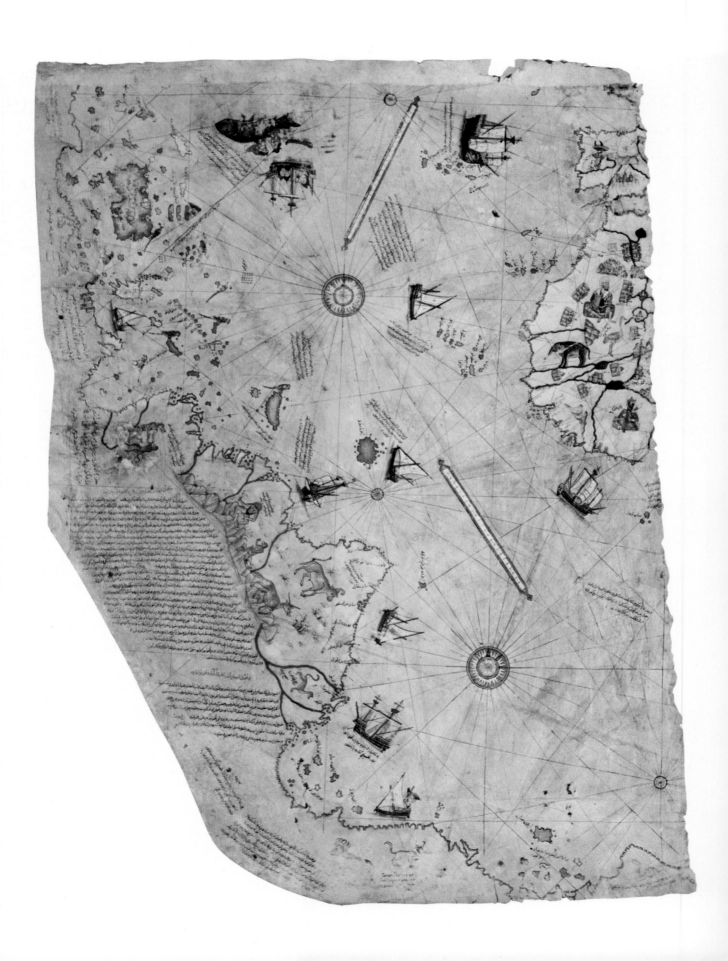

Opposite: the Piri Re'is map, said to have belonged to a Turkish Navy admiral, discovered in the Topkapi Seraglio in 1935—not in the 18th century as von Däniken apparently believes. According to von Däniken this map shows mountain ranges of Antarctica now so deeply covered with snow that it requires echo-sounding apparatus to map them. He also says that this is an aerial view, which required the help of space people in aircraft.

A Map From Outer Space?

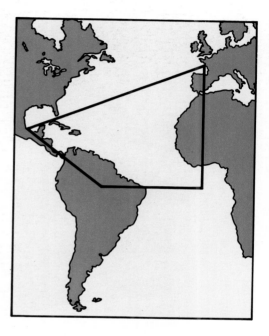

would defend his rather reckless methods of research and presentation, and if his hypothesis is correct, it will require far more proof. Yet it has to be admitted that the possibility that Earth has been visited by advanced beings is a real one. There could be many inhabited planets where life developed much earlier than it has on this planet. If their technological advances occurred at a similar rate to ours but a few thousand years earlier, they may well have solved the problems of traversing vast expanses of the Universe. If so, they could have left some evidence of their visit, just as we from Earth have done on our own uninhabited satellite. Perhaps in fact the space visitors have never left us but have kept their distance and monitor our progress from the skies. Could it be that we are no more than an intergalactic breeding experiment?

Whatever the physical proof—or lack of it—as regards space visitors and human origins, the record shows that strange objects have been reported in our skies since the earliest recorded history. Until a logical explanation can be offered for these— and their modern equivalent the flying saucers—we should probably keep an open mind about Erich von Däniken's sensational theories.

Above: the black line indicates the area that the Piri Re'is map covers as Dr Crown sees it. The coasts of South America accurately mapped were already known to cartographers in 1513, when the map was drawn. Antarctica, ice-covered or not, does not appear on Dr Crown's projection.
Left: an azimuthal projection based on Cairo and proposed by von Däniken as the one used in the Piri Re'is map. Azimuths project the curved surface of a globe onto a flat surface based on a given center, so distortions always occur. However, Dr A. D. Crown of the University of Sydney points out that the distortion cited by von Däniken to explain the peculiar length of South America on the Piri Re'is map would not occur in an azimuthal projection that was based on Cairo.

Chapter 9
Spacecraft in Early Times

We consider generally that UFO sightings began with Kenneth Arnold's Idaho "flying saucers" in 1947, but there is a long history of strange objects in the sky. The earliest photographs of UFOs were apparently taken in 1883 by a Mexican astronomer. But the early sightings are worldwide. Have we been under observation for generations? Some scholars even suggest that the Bible may report UFO activity—was the famous vision of the Old Testament prophet Ezekiel really a description of his encounter with a spaceship and its occupants? And when he later says he was lifted and taken away by the spirit, is he saying he was transported by a spacecraft?

Over a thousand strange objects were reported in the skies above Sweden in a 12-month period. That is remarkable in itself. What makes it more remarkable is that it occurred in 1946—the year before flying saucers hit the world headlines. It could not, therefore, be explained away as mass hysteria, though the Swedish press did take up the story and the objects became known as "ghost rockets." They were usually seen at night, and some accounts describe them as cigar-shaped with orange flames issuing from the tail. In the period from July 9 to July 30 alone, the Swedish military received more than 600 reports, mostly of fast-moving colored lights in the night sky.

These UFOs caused serious concern at the time in both the United States and Scandinavia. It was feared that German scientists in the Soviet-occupied zone of Germany were developing and testing a new secret weapon. Even before this, during World War II, there had been reports of UFOs seen mainly at night. Each side had found them perplexing, and each had thought that they were secret weapons being developed by the other. They were given the name of "foo fighters" by British and American pilots who encountered them. What impressed airmen was that these small colored lights appeared to be under intelligent control.

A spate of reports of foo fighters started to come in about November 1944. They were encountered by night fighters flying over the Rhine north of Strasbourg. Typical of the accounts is this report filed by Lieutenant David McFalls, and

Opposite: spaceships in ancient Egypt—a pre-von Däniken view from an American science fiction magazine published in 1948. Strange flying objects were reported long before Arnold saw his flying saucers in 1947, and the witnesses form an oddly assorted crew—from Old Testament prophets to a Kansas farmer. They even include Christopher Columbus as he sailed the Atlantic Ocean.

UFOs in the 19th Century

Below: an illustration for Jules Verne's *The Clipper of the Clouds* (1886). Robur, the hero of the story, has a shadowy background and unimaginable wealth. He builds a giant airship, which takes him around the world and into a series of incredible adventures. Using the technology of his age, Verne clearly visualized Robur's craft—the *aeronef* —as a combination of an airship and a seagoing vessel.

published in Harold Wilkins' book *Flying Saucers from the Moon.*

"At 0600 near Hagenau, at 10,000 feet altitude, two very bright lights climbed toward us from the ground. They leveled off and stayed on the tail of our plane. They were huge, bright orange lights. They stayed there for two minutes. On my tail all the time. They were under perfect control. Then they turned away from us, and the fire seemed to go out."

Others reported even more incredible encounters. The foo fighters would sometimes race ahead or fly abreast of aircraft, and then drop behind and follow the tail. When the pilots made abrupt changes of course, the balls of fire remained on their tails, in one case for as long as 75 minutes. This was wartime, so the aircraft showed no lights. The foo fighters were certainly not reflections of the planes.

It is understandable that these UFOs and the ghost rockets seen over Sweden two years later should have been regarded as a secret weapon developed by the enemy. But history has shown that they did not belong to either side. History also shows that UFOs have been with us for as long as man has been able to record his past, and probably before then as well. As a result "ufologists"—the popular name for UFO researchers—have plundered the history books and produced a profusion of volumes with titles like *The Bible and Flying Saucers*, *God Drives a Flying Saucer*, and *Flying Saucers Through the Ages.*

If modern science has been unable to prove the reality of UFOs with radar and photographic evidence, then clearly a rummage through ancient records is not going to provide proof either. The witnesses are no longer around to be questioned. The statements they made, or that are attributed to them, are usually highly colored but insufficiently detailed to enable us to make a positive identification at this late stage. In most cases, however, we can rule out satellites, aircraft, or the other products of modern technology that feature so prominently today as the explanation for many UFOs.

One of the most remarkable early UFO periods was 1896–7, when craft resembling airships were seen by many people in different parts of the United States. They were usually described as cigar-shaped, and often the reports spoke of powerful searchlights aimed at the ground from these aerial visitors. Most witnesses assumed they were man-made, but although airships were certainly at the design stage, there is no evidence that any were yet in operation at that time.

The 19th-century sky visitors provoked startling headlines, and with good reason. For example, a Kansas farmer, Alexander Hamilton, reported an encounter that would make front page news in any newspaper even today. He was awakened by the noise of his cattle at 10:30 p.m. on April 19, 1897, and on going to investigate he saw "an airship slowly descending upon my cow lot about 40 rods from the house." He called his son and a tenant, and the three men rushed outside with axes. The airship was now hovering just 30 feet above the ground. It was cigar-shaped, about 300 feet long, and had a transparent, brilliantly lit undercarriage. In a sworn statement Hamilton

reported that the airship carried "six of the strangest beings I ever saw." They turned a beam of light directly on him. Then a large turbine wheel, about 30 feet in diameter, began to buzz as it revolved slowly "and the vessel rose as lightly as a bird."

When the airship reached an estimated height of 300 feet it stopped and hovered. The three men saw that a thick cable had been put around a two-year-old heifer's neck, and the occupants began to hoist it aboard. The airship then flew off as the witnesses watched in amazement. Next morning Hamilton could not believe what he had seen, and he went in search of the animal. He learned that a neighboring farmer had found the hide, head, and legs of the butchered heifer in his field.

Hamilton ended his statement with these words: "I don't know whether they are devils or angels, or what; but we all saw them, and my whole family saw the ship, and I don't want any more to do with them."

The *Colony Free Press* of Kansas soon had this explanation for its readers: "The *Free Press* . . . is *now* of the opinion that the airship is not of *this world*, but is probably operated by a party of scientists from the planet Mars, who are out, either on a lark, or a tour of inspection of the solar system in the cause of science."

So the Martians were thought to be responsible even in the 19th century!

Above: a photograph of the first British airship for military purposes, shown under test in 1907. It is at anchor near Farnborough, England. In the beginning of the 20th century the airship was the most technologically advanced product of aeronautics. Perhaps not surprisingly, the scattered reports of UFO sightings during that period most often described the objects as similar to airships.

Did Columbus See a UFO?

Right: dark balls in the sky, shown in a
16th-century woodcut. They were seen in
Basel, Switzerland in August 1566, and
resembled the blood-red round objects
sighted in Nürnberg four years previously.

José Bonilla, a Mexican astronomer, enjoyed a UFO show
all his own in August 1883. He was taking photographs of
sunspot activity at the Observatory of Zacatecas when he
noticed strange objects crossing the face of the sun. In the
two hours before clouds obscured his view he counted 283.
The objects, traveling east to west, were apparently luminous.
Bonilla succeeded in photographing some of the objects.
These were probably the first UFO photographs ever taken,
but like modern versions, they did nothing to help identify the
objects. Other observatories are said to have seen the objects as
well. This would indicate that they were outside the earth's
atmosphere. When Camille Flammarion, renowned French
astronomer, was shown the photographs he tentatively suggested that they were insects, birds, or dust.

The 19th century began with an impressive sighting at Baton
Rouge, Louisiana. On the night of April 5, 1800, an object
described as "the size of a large house" passed about 200 yards
above the ground. It was brightly lit, caused heat to be felt on
the ground, and took 15 minutes to disappear to the northeast.

The English astronomer Edmund Halley was sent a UFO
report by an Italian professor of mathematics. He reported
seeing in March 1676 a "vast body apparently bigger than the
moon," which crossed over all Italy. It was at an estimated
height of 40 miles, made a hissing sound and a noise "like the
rattling of a great cart over stones." The professor calculated
its speed at 160 miles a minute—9600 miles per hour. Halley's
comment was: "I find it one of the hardest things to account
for, that I have ever yet met." The next year, in 1677, Halley
himself reported observing "a great light in the sky all over
southern England, many miles high."

A number of years later in 1716 Halley saw a UFO that
illuminated the sky for more than two hours. He could read a
printed text in the light of this object. After two hours the
brightness of the object was reactivated "as if new fuel had

been cast in a fire."

In 1561 many people in Nürnberg reported seeing blood-red, bluish, or black balls and circular disks in the neighborhood of the rising sun. They were visible for an hour, and then appeared to fall to the ground as if on fire.

Even Christopher Columbus saw a UFO. He was patrolling the deck of the Santa Maria at about 10 p.m. on October 11, 1492 when he saw "a light glimmering at a great distance." He summoned another member of the expedition who also saw the light. It vanished and reappeared several times.

Some of these UFOs may well be accounted for by normal events. Others would probably stand up to rigorous study by today's scientists and join the large number of unsolved riddles on record. A few—like the aerial craft reported to have hovered over Bristol, England around 1270—would be too fantastic to study. The craft caught an anchor in a church steeple, and an occupant scampered down a ladder to free it. He was stoned to death by the crowd, which, regarding him as a demon, burned his body.

Shining Wheel in the Sky

As a British steamer plowed its way through the Persian Gulf near Oman in the summer of 1906, an enormous wheel of light appeared. The vast wheel, seemingly bigger than the ship itself, was revolving in the sky not far above the surface of the water at that point. Vivid shafts of light emitted from the huge wheel and passed right through the steamer. But these beams fortunately did not interfere with the functioning of the boat in any way.

Since 1760 seamen have recounted sightings of unidentified flying objects in the form of a wheel. The Persian Gulf sighting if 1906 was one of eleven recorded reports between 1848 and 1910. Like most of the sea accounts of mysterious luminous wheels, this one remarked on the eery silence of the phenomenon. Also in common with most other such reports, nothing was said about humans or humanlike beings in the wheels, even though the ascent and descent of these objects were obviously controlled.

Were such glowing wheels in the sky an early and less sophisticated form of flying saucer? Were they operated by beings from other planets who kept themselves hidden or were invisible? Were they just visions of mariners too long at sea? No one has found an answer.

Left: Christopher Columbus on the deck of the *Santa Maria*. On the night before the historic sighting of the New World, Columbus saw a glimmering light at a distance. He called one of his crew members to observe it with him. The light vanished and reappeared several times as they watched during the night.

The story is reminiscent of an old Irish legend of which there are a number of versions. According to one account, a marvel happened one Sunday while the people were at mass in the borough of Cloera. "It befell that an anchor was dropped from the sky, with a rope attached to it, and one of the flukes caught in the arch above the church door. The people rushed out of the church and saw in the sky a ship with men on board, floating before the anchor-cable, and they saw a man leap overboard and jump down to the anchor, as if to release it. He looked as if he were swimming in water. The folk rushed up and tried to seize him; but the priest forbade people to hold the man, for it might kill him, he said. The man was freed, and hurried up to the ship, where the crew cut the rope and the ship sailed away out of sight. But the anchor is in the church, and has been ever since, as a testimony."

An even more remarkable account of contact between earth people and space visitors is said to have taken place some time during the reign of Charlemagne, who lived from A.D. 742 to A.D. 814. According to the version published by Brinsley le Poer Trench in his book, *The Flying Saucer Story*, spacecraft took away some of Earth's inhabitants to show them something of the way of life of space people. But when they returned the populace were convinced that their fellow men were members of the space race, whom they regarded as sorcerers. They seized these people and tortured them, and put many to death. Le Poer Trench quotes this passage from the original:

Right: a 16th-century Persian manuscript showing the legendary Kai Ka'ns going up in a fantastic flying machine. It is powered by enticing tied-up eagles to reach for raw meat impaled at the top.

"One day, among other instances, it chanced at Lyons that three men and a woman were seen descending from these aerial ships. The entire city gathered about them, crying out that they were magicians and were sent by Grimaldus, Duke of Beneventum, Charlemagne's enemy, to destroy the French harvests. In vain the four innocents sought to vindicate themselves by saying that they were their own countryfolk, and had been carried away a short time since by miraculous men who had shown them unheard of marvels, and had desired them to give an account of what they had seen."

These four were about to be thrown into a fire when they were saved by Agobard, Bishop of Lyons. He listened to the accusations and to the defense, and decided that the four had not fallen from the sky at all. They were set free.

The further back in time we go, the more astonishing are the stories that are uncovered. The ancient Indian Sanskrit texts tell of gods who fought in the sky on aircraft—*vimanas*—equipped with deadly weapons. Many scholars regard these accounts as no more than the vivid imaginings of simple people who wished to bestow upon their gods the greatest conceivable powers. However, modern writers have pointed to some astonishing similarities between these stories and 20th-century inventions. The following is from Protap Chandra Roy's translation of the ancient Indian manuscript the *Drona Parva* which he did in 1889, and which is quoted in a book by Desmond Leslie and George Adamski called *Flying Saucers*

Gods and Their Flying Machines

Below: an Indian manuscript illustration of flying deities. In it the gods are coming to observe the outcome of one of the battles of the mythical champion Rajah Karna, told in the *Mahabharata*. Von Däniken has cited the numerous references to flying gods and flying machines—*vimanas*—as evidence for his theories of visitors from space.

Biblical UFOs

Below: a 16th-century painting of the *Adoration of the Magi*. The peculiar behavior of the star that moved before the Three Wise Men in guiding them to Bethlehem, and came "to rest over the place where the child was" has made ufologists speculate that the star was in fact a spacecraft. It has also been pointed out that the Three Wise Men themselves—being dark and foreign—bear a remarkable resemblance to the mysterious three men in black of recent UFO history.

Have Landed:

"A blazing missile possessed of the radiance of smokeless fire was discharged. A thick gloom suddenly encompassed the hosts. All points of the compass were suddenly enveloped in darkness. Evil-bearing winds began to blow. Clouds roared into the higher air, showing blood. The very elements seemed confused. The sun appeared to spin round. The world, scorched by the heat of that weapon, seemed to be in a fever. Elephants, scorched by the energy of that weapon, ran in terror, seeking protection from its terrible force. The very water being heated, the creatures who live in the water seemed to burn. The enemy fell like trees that are burned down in a raging fire . . . The steeds and the chariots, burned by the energy of that weapon, resembled the stumps of trees that have been consumed in a forest conflagration. Thousands of chariots fell down on all sides. Darkness then hid the entire army. . . ."

The scripts tell of a number of deadly weapons, including *Indra's Dart*, which was operated by a circular reflecting mechanism. It did not fire weapons, but was switched on like a searchlight. The shaft of light that was produced was aimed at a target which it was able to consume with its power.

These descriptions could easily be applied to nuclear bombs and laser beam weapons. Who were the possessors of such inventions, and what were the aircraft in which they traveled? W. Raymond Drake, in his book *Gods and Spacemen in the Ancient East*, quotes from a Sanskrit text which explains "the art of manufacturing various types of Aeroplanes of smooth and comfortable travel in the sky, as a unifying force for the Universe, contributive to the well-being of mankind." It gives a definition of a vimana as "that which can go by its own force like a bird, on earth, or water, or in air." Today, that which can travel in the sky from place to place, land to land, or globe to globe, is called "vimana" by scientists in aeronautics.

Some of the contents of the manuscript are listed as: "The secret of constructing aeroplanes, which will not break, which cannot be cut, will not catch fire, and cannot be destroyed. The secret of making planes motionless. The secret of making planes invisible. The secret of hearing conversations and other sounds in enemy planes. The secret of receiving photographs of the interior of enemy planes. The secret of ascertaining the direction of enemy planes' approach. The secret of making persons in enemy planes lose consciousness. The secret of destroying enemy planes . . .

"Metals suitable for aeroplanes, light and heat-absorbing, are of 16 kinds . . .Great sages have declared that these 16 metals alone are the best for aeroplane construction."

This translation, done by Maharishi Bharadwaja, is called *Aeronautics* and is described as *A Manuscript from the Prehistoric Past*. It is published by the International Academy of Sanskrit Research, Mysore, India. The unresolved question is: Are scripts like these—possibly as old as the Old Testament—the work of inspired science fiction writers of their day, or are they eyewitness accounts of incredible events?

Perhaps further research will throw light on these perplexing questions. In the meantime, the UFO has enabled imaginative writers to reassess well-known past events in a new light. The Bible has proved to be a rich source of material. The star of Bethlehem that guided the Three Wise Men to the birthplace of Jesus is regarded by some ufologists as a flying saucer. They point to the fact that the wise men followed it until it "came to rest over the place where the child was." If it had been a bright star or planet, it would not have stopped at any point. But an unidentified flying object might well have hovered above one place. A writer who has explored UFO links in the Bible is Barry L. Downing, pastor of Northminster Presbyterian Church in Endwell, New York. He concludes that the miracles of the Old and New Testaments, usually ascribed to supernatural forces, could as reasonably be given an extraterrestrial interpretation. Instead of being spirits or messengers of God, angels could have been space visitors. God, reported to have spoken, could have been a being from another planet endeavoring to guide Earth people during crises.

Examining *Exodus* he points out that when the Israelites fled from Egypt, the Lord is said to have led them "by day in a pillar of cloud . . . and by night in a pillar of fire to give them light that they might travel by day and night . . ." This pillar, says Downing, was a flying saucer. Later, during their years in the wilderness, according to Downing it is a UFO that is referred to variously as the "Lord," "the angel of God," and a "cloud." Certain verses in *Exodus*, says Downing, clearly show that Moses was in contact with space visitors who gave him the Ten Commandments, for example: "Then Moses went up

Above: Moses receiving the Ten Commandments on Mount Sinai. B. L. Downing, a pastor of New York, considers that the "thick cloud" into which Moses vanished must have been a heavenly UFO.
Below: a painting of *Moses and the Burning Bush*. Downing believes that the bush was burning because a flame-belching UFO had set down in the middle of it.

Ezekiel's Vision of a Spaceship?

The Book of Ezekiel in the Old Testament contains the prophet's description of his vision. It is open to many interpretations, but was long accepted purely as the mystical experience of a visionary. Today, however, some commentators explain it in terms of 20th-century technology. These three illustrations show different interpretations of the biblical prophet's strange vision. Right: a traditional representation. It shows a conventional chariot with four wheels, and four winged creatures having the faces of a man, a lion, an ox, and an eagle exactly as described by Ezekiel.

Below: another view of the vision of the prophet. This artist has painstakingly worked out a literal interpretation of the account on the basis that it was miraculous and showed the power of the Lord.

on the mountain, and the cloud covered the mountain. The glory of the Lord settled on Mount Sinai, and the cloud covered it six days; and on the seventh day he called to Moses out of the midst of the cloud. Now the appearance of the glory of the Lord was like a devouring fire on the top of the mountain in the sight of the people of Israel. And Moses entered the cloud." Downing concludes: "Moses seems to have gone aboard the UFO, at which time he received stone tablets from the being in the UFO."

Most Christians will be excused for preferring a more orthodox, or even supernatural view of religious happenings. Few scholars would accept Downing's theories either. Like others before him with a new interpretation of our past, he tends to overstate his case to be persuasive, and risks losing credibility. But there are others who believe the Bible contains important UFO material, viewed from a scientific rather than a religious viewpoint. The Biblical account that has attracted the greatest attention among such believers is the striking experience of the Old Testament prophet Ezekiel.

The Book of *Ezekiel* begins with the remarkable vision that has caused so much excitement among modern ufologists. It is helpful to know the essential ingredients of Ezekiel's story before looking at the modern interpretation. It says: "Now it came to pass in the thirtieth year, in the fourth month, in the fifth day of the month, as I was among the captives by the river of Chebar, that the heavens opened and I saw visions of God . . . And I looked, and, behold, a whirlwind came out of the north a great cloud, and a fire infolding itself, and a brightness was about it . . . And out of the midst thereof came the likeness of four living creatures. And this was their appearance; they had the likeness of a man. And every one had four faces, and every one had four wings. And their feet were straight feet; and the sole of their feet were like the sole of a calf's foot; and they sparkled like the color of burnished brass. And they had the hands of a man under their wings on the four sides . . . Their wings were joined one to another; they turned not when they went; they went every one straight forward . . . two wings of every one were joined one to another, and two covered their bodies . . ."

Ezekiel goes on to describe how each of the four "living creatures" had an identical shining wheel that accompanied them even when they rose from the ground. Two things were strange about the wheels. They were constructed as though one wheel was inside another, and they could move in any direction without turning. Ezekiel gives many other details of his experience that has usually been regarded as visionary—simply as another attempt to describe mystical experiences in a way that others can understand. But it is different from other biblical visions because Ezekiel, a priest, goes into surprising detail. Whatever his belief about the Lord and angels, Ezekiel appears to have experienced something so totally alien to his knowledge and understanding that he sets out to give as much detail about the *construction* of what he saw as possible. Believers in flying saucers were quick to interpret his vision as a description of a spaceship and its occupants. Their explanation, however, left as many unanswered questions as the belief that Ezekiel's experience was purely mystical.

Erich von Däniken is among the writers who have postulated an extraterrestrial explanation of Ezekiel's vision. When Josef F. Blumrich, who had spent the greater part of his life on the

Below: this interpretation of Ezekiel's vision is based on present-day knowledge of space travel and astronauts. It was painted by a British aeronautical artist and pilot, Keith Mosely, who felt that von Däniken's theory was worth some thought. The four faces depicted literally in the first interpretation are in this case shown as three big insignia on the arms and chest of an extraterrestrial being, plus his own face. The machine that is propelled by rotor blades and has a base like a wheel carries the space visitors to the ground from their hovering mother ship. This fits in with Ezekiel's description of the men having wings, and of "a wheel upon the earth beside the living creatures, one for each four of them."

Modern View of the Prophet's Chariot

design and analysis of aircraft and rockets, read the suggestion he set about disproving it. After painstaking research, Blumrich decided that Ezekiel did see a spaceship. What is more he has produced detailed drawings of its construction, based on the prophet's description, which he says is a little ahead of our present technology. He gives the full details in his book, *The Spaceships of Ezekiel*. "Seldom has a total defeat been so rewarding, so fascinating, and so delightful!" he declared.

Blumrich, chief of the systems layout branch of the National Aeronautical and Space Administration (NASA), has worked on spaceship construction since he went to the United States from his native Austria in 1959. He is cobuilder of the Saturn V, and holder of many patents. In fact, Blumrich had been involved in designing and testing the wheel mechanism so accurately described by Ezekiel. This made him realize that the Bible texts could be describing a technical device rather than an ethereal vision. His further investigation led him to conclude that Ezekiel must have possessed extraordinary gifts of observation and an almost photographic memory. "This made it possible not only to develop a simple sketch, but also to express dimensions, weights, and capabilities in figures."

The four living creatures described by the prophet were not living at all, says Blumrich. They were the four legs of the huge spaceship, each with a wheel that could move forward and backward or revolve sideways. Such a wheel is now technically feasible and patented. Their wings were rotor blades. The geometry of these helicopter units made it appear that they touched each other when they rotated. The main propulsion system was by rocket engine, with the propellant stored in the huge top-shaped central body. The helicopter blades would have been used only for final touchdown maneuvers.

Blumrich believes that the results of his research "show us a space vehicle which beyond any doubt is not only technically feasible but in fact is very well designed to fulfill its functions and purpose. We are surprised to discover a technology that is in no way fantastic but even in its extreme aspects, lies almost within reach of our own capabilities of today, and which is therefore only slightly advanced beyond the present state of our technology. Moreover, the results indicate a spaceship operating with a mother spacecraft orbiting the earth.

"What remains fantastic is that such a spacecraft was a tangible reality more than 2500 years ago!"

The NASA scientist points out that later in his account the prophet describes being lifted and taken away by the spirit. This, he suggests, is further evidence that the vision was a spacecraft that transported Ezekiel to various places during the four times he met the extraterrestrial visitors.

Even though some puzzling biblical texts have provided the basis for the design of a space vehicle technically superior to our present designs, Blumrich's argument does not provide the indisputable proof required to convince the world that visitors from outer space have been in touch with mankind at points during its evolution. But a glance back through a few thousand years of our history is enough to show that UFOs—whatever they may be—are not a 20th-century phenomenon.

Below: Josef F. Blumrich, a NASA space engineer who spent most of his life in designing and building aircraft and rockets—among them the giant Saturn V rocket. He became fascinated with von Däniken's view of the vision of Ezekiel as the visit of a UFO, and plunged into research to refute him. Instead he found that he had become a convert. He worked out a reconstruction of the spacecraft he thought Ezekiel had witnessed, even including details of how it might have operated.

Blumrich's version of what Ezekiel may have been trying to describe.
Right: the spacecraft as it would appear upon entering the Earth's atmosphere.
The four "men" that Ezekiel thought he saw were, says Blumrich, really four
legs of the spaceship, which act as a kind of helicopter. During entry into the
atmosphere, they fold back over the top of the spaceship, so making the
braking potential fully effective. This phase of the flight would have been
completed at the edge of the atmosphere, before Ezekiel saw the spacecraft.

Below: the landed spacecraft standing on the four helicopter legs. The brief
bursts of the control rockets as the craft searched for a suitable landing space
were interpreted by Ezekiel as lightning, and he saw the radiator of the reactor
glowing like burning coals.

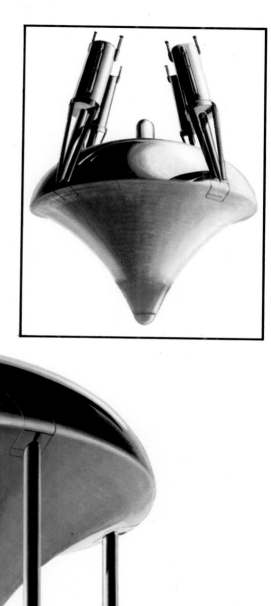

Chapter 10
The First of the Contacts

If there are mysterious spacecraft observing earthly activity, do the spacemen presumably flying the airships ever try to make contact with human beings? A varied and colorful group of witnesses claim that they have done so. Here are some of the amazing reports of contacts made with men—and women!—from outer space, whose curiosity about the human race appears to be scientific and methodical. Have individuals been kidnaped by spacemen for intensive examination? Are human beings being studied like laboratory animals by creatures from some other world in the Universe? Can we believe the reports of the people who say they have been inside a UFO?

The meeting in the California desert between George Adamski and a man with long blond hair deserves to go into the history books as one of the most extraordinary events ever reported. If Adamski was telling the truth, his encounter near Desert Center in 1952 was a turning point in our history: the moment when an inhabitant of Earth came face to face with a visitor from Venus. If it was a hoax, it was one of the most elaborate and successful ever staged. As a result of it Adamski, an amateur astronomer, was able to travel widely and lecture about his contacts with extraterrestrials, and was feted by royalty.

The modern flying saucer era was just over five years old at the time Adamski claimed he met a Venusian. Many people had become convinced that the flying saucers were interplanetary space vehicles, but the absence of any apparent attempt to contact the Earth's inhabitants was puzzling. Rumors of tiny creatures who had been taken from the scene of UFO crashes—dead or alive—were rife. Official secrecy was always given as the reason for absence of detail in such cases.

George Adamski, a 61-year-old Polish-born American, lived in the shadow of the famous Mount Palomar Hale Observatory in California. He not only believed that UFOs were real, but he also felt sure that whoever was piloting them were humanlike beings. He and a group of fellow believers had been told of a number of flying saucer incidents in the area in which the objects were seen to descend close to the ground and even to land. The

Opposite: this drawing of a space visitor was drawn from a description given by two Americans, Betty and Barney Hill, who sighted a UFO and later could not account for the following two hours. It was under hypnosis treatment that they recalled their experience in detail, and this and other drawings were based on their recollection.

Above: George Adamski on a lecture tour in Europe in 1959, pictured with the secretary of the Dutch UFO Contact Group. In 1952 Adamski had claimed that he had met and talked with Venusians.
Below: Adamski's snapshot of the object he said was a Venusian scout ship in takeoff (arrowed). This was his first meeting with Venusians on November 20, 1952. It was witnessed by a group of six of his friends who were interested in investigating UFOs.

Adamski group decided not only to try to get good close-up photographs of the saucers but also to try to contact their occupants.

It was with this object in mind that Adamski and six others went into the California desert on November 20, 1952. Adamski felt he was somehow being guided by the space visitors, and the others allowed him to lead the way. They drove to Desert Center and then for another 10 miles in the direction of Parker, Arizona. Soon after stopping on the roadside to eat, the saucer hunters were rewarded by the sight of a large cigar-shaped UFO that had also attracted the attention of military planes in the area.

Adamski, a nondriver, shouted: "Someone take me down the road—quick! That ship has come looking for me and I don't want to keep them waiting! Maybe the saucer is already up there somewhere—afraid to come down here where too many people would see them." Two of the party drove off with him at once.

The group of saucer enthusiasts had already decided on the basis of personal observations that the cigar-shaped UFOs were mother ships: huge carriers that acted as bases for the smaller disk-shaped saucers. They did not, therefore, expect the object they were watching to make a landing, and when Adamski reached a desired spot between half-a-mile and a mile from the picnic area, he and the others in the car were not surprised to see the cigar-shaped UFO disappear over the nearby mountains. Adamski impored his two friends to leave him with his six-inch telescope, tripod and photographic equipment, and they went back to the rest of the party.

Within minutes his attention was attracted by a flash in the

sky, and he saw a small craft that seemed to drift toward him. It settled itself in a cove about half-a-mile away. Adamski immediately found the flying saucer in the telescope and began taking photographs through a camera attachment. After taking several photographs Adamski removed the camera and placed it in its box. He then took out a Kodak Brownie in order to try a long shot. As he did so a couple of planes roared overhead and the saucer dropped down out of sight. He assumed it had returned to its mother ship, but after a few minutes he suddenly saw the figure of a man standing at the entrance of a ravine some way away. The man beckoned to him. Puzzled, Adamski decided that someone needed help, and he set off in that direction.

As he got closer Adamski noticed that the figure was slightly smaller and younger than he was, and dressed in ski-type trousers. His long hair reached to his shoulders, which was most unusual for a man at that time.

"Suddenly, as though a veil was removed from my mind," Adamski wrote later, "the feeling of caution left me so completely that I was no longer aware of my friends or whether they were observing me as they had been told to do. By this time we were quite close. He took four steps toward me, bringing us within arm's length of each other.

"Now, for the first time I fully realized that I was in the presence of a man from space—*a human being from another world!*"

Adamski estimated the man's age as about 28, his height as 5 feet 6 inches, and his weight around 135 pounds.

"He was round-faced with an extremely high forehead; large, but calm, gray-green eyes, slightly aslant at the outer corners; with slightly higher cheekbones than an Occidental, but not so high as an Indian or an Oriental; a finely chiseled nose, not conspicuously large; and an average size mouth with beautiful white teeth that shone when he smiled or spoke."

Conversation was difficult, for the spaceman did not appear to speak much English and Adamski did not recognize the man's language. Instead, they communicated by a combination of telepathy and sign language. In this way Adamski established that the man came from Venus, that the space visitors were concerned about bombs and radiations emanating from the Earth, and that they operated their craft by the law of attraction and repulsion. During this discussion the Venusian pointed to his flying saucer, which was hovering above the ground unnoticed by Adamski. When it was time for him to depart the man from Venus walked over to his spacecraft with Adamski. He indicated that the Earthman should not get too close, but Adamski forgot and moved underneath the outer flange of the hovering saucer. "My arm was jerked up, and almost at the same instant thrown down against my body," he said. "The force was so strong that, although I could still move the arm, I had no feeling in it as I stepped clear of the ship."

The Venusian assured Adamski that feeling would return to his arm, which it did after some weeks. What concerned Adamski more was that the seven exposed photographic plates in his right pocket may have been affected by the strange power. He put his hand in his pocket and removed them. The Venusian indicated

Adamski and the Venusian Visitor

Above: a Venusian's footprint that Adamski claims was deliberately left by the space visitors. There were three sets of deep and distinct prints, and Adamski believed the symbols contained a special message for the people on Earth.
Below: sketches of the patterns found on the Venusian footprints made by Betty Bailey, one of the witnesses.

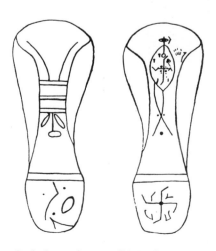

Left footprint   **Right footprint**

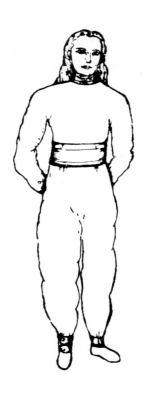

Above: a sketch of the Venusian made by Alice Wells who watched the meeting through binoculars. Adamski remarked that the sketch "conveys the broad features of his appearance but is far short of doing him justice." Mrs. Wells was one of the witnesses who signed a notarized statement.

that he would like to take one. He made it clear that he intended to return it at a future date.

The time had now arrived for the brief meeting between two civilizations to end, and the Venusian bade his Earth companion farewell. He stepped onto a bank behind the saucer, from there onto its flange, and entered his spacecraft without Adamski seeing how he did so. In moments the flying saucer, looking like a heavy glass bell, rose up and moved off into space.

What made Adamski's account seem plausible to many people was that his six friends had witnessed the encounter from a distance. His book *Flying Saucers Have Landed*, which was co-authored with the British writer Desmond Leslie, contains photographs of their testimonies, signed before notaries public. They said:

"We, the undersigned, do solemnly state that we have read the account herein of the personal contact between George Adamski and a man from another world, brought here in his Flying Saucer "Scout" ship. And that we were a party to, and witnesses to the event as herein recounted."

The declarations were signed by Mr. and Mrs. Al C. Bailey of Winslow, Arizona; Dr. & Mrs. George Hunt Williamson of Prescott, Arizona; Mrs. Alice K. Wells, owner of Palomar Gardens and operator of the café there; and Mrs. Lucy McKinnis, Adamski's secretary.

They had not been able to hear the conversation, of course, but they had seen the small domed saucer, and had watched Adamski speaking to a man in a one-piece suit. Mrs. Wells even drew a picture of the Venusian after looking at him through binoculars.

Before he left the space visitor drew Adamski's attention to the footprints he had left in the sand. He was wearing shoes decorated with an inscription on the soles, designed, it seems, to leave a message for the world. Williamson took a number of plaster casts of the footprints immediately after the strange desert encounter. He has since written a number of books on flying saucers which include a detailed analysis of what these inscriptions may represent.

On December 13, 1952 the Venusian scout ship visited Adamski again. This time there was no contact, but as it swooped low over Palomar a porthole window opened and Adamski's photographic plate was thrown out. A hand waved and the craft sped away, but not before Adamski had taken a number of pictures through his telescope. These showed the craft in great detail, and established the Adamski saucers as the classic shape in the mind of the public. His is an upturned saucer with three balls underneath, appearing to be landing gear. On top is a circular cabin with portholes. Many other observers have claimed to have seen similar flying saucers and to have photographed them, but the Adamski pictures were the most influential and widely accepted among those who believed in UFO contacts.

Adamski's later encounters with his friends from other worlds were even more fantastic. In his second book *Inside the Space Ships*, Adamski claimed, following a hunch, he went to Los Angeles and he booked into his favorite hotel. There he was approached by two men who greeted him with a special handshake identifying them as space dwellers. They looked like ordinary people, had on normal clothes and wore their hair short, and spoke with hardly a trace of an accent. They drove him out of the city for a rendezvous with the scout ship. He learned during this journey that one of his companions was from Mars, the other from Saturn. When they reached the flying saucer Adamski was delighted to find his Venusian friend. This time, for some inexplicable reason, he spoke perfect English.

The men from the three planets took him aboard the craft and for a ride in the flying saucer. Adamski gave them names to help identify them in his narrative, although he said they did not use names as we do. The Venusian he called Orthon, the Martian Firkon, and the Saturnian Ramu.

A lens in the floor of the saucer enabled Adamski to view the earth as the scout ship climbed into space. It went aboard the mother ship where Adamski was greeted by other extra-terrestrials including two extremely beautiful women pilots who kissed his cheek. It was during this space trip that Adamski was told the reason for the coming of the saucers. It was to warn "of the grave danger which threatens men of Earth today" through

He Meets Other Extraterrestrials

Below: an artist's conception of the meeting of Adamski and the Venusian in the California desert.

A Description of Travel in Space

atomic explosions. There was also a danger that atomic explosions on earth could upset the balance of the galaxy and the lives of others on neighboring planets.

During another space flight with his friendly planetary contacts—whose ages ranged from 40 to several hundred years—Adamski claims he was taken to view the moon. His space friends told him that earth scientists were wrong about the satellite. Its conditions were not as extreme as imagined.

"There is a beautiful strip or section around the center of the moon in which vegetation, trees, and animals thrive, and in which people live in comfort. Even you of Earth could live on that part of the moon, for the human body is the most adaptable machine in the Universe."

On the side facing the earth, vegetation was sparse but

Right: the photograph that Adamski said had been substituted for one of his by the Venusians. They had borrowed one of his photographic plates on their first meeting with him, and returned it on December 13 by dropping it from their scout ship near the spot he had set up his telescope. The original subject had been washed off. Adamski said the substitute picture was of writing from another planet.

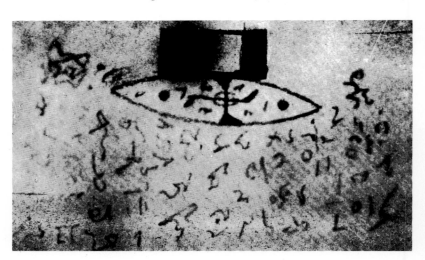

Right: a telescopic picture of the underside of the Venusian spacecraft, taken by Adamski as it dropped off the photograph.

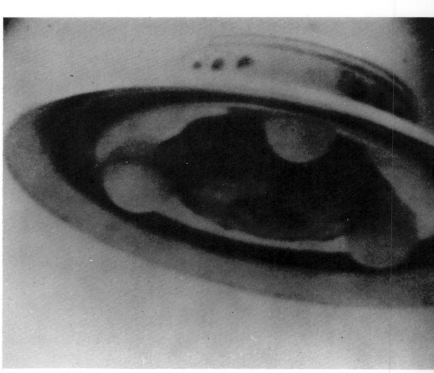

Adamski saw a small furry four-legged animal running about. Later, when he was taken to the other side of the moon, Adamski claims he saw cities, mountains, rivers, and lakes. "Actually, we seemed to be drifting over the rooftops, and I could see people walking along clean, narrow streets." Adamski was also shown live pictures from Venus, beamed onto an invisible screen. These too showed cities and earthlike scenery. He noticed that animal life was similar to ours, and he saw birds like canaries as well as horses and cows.

George Adamski died on April 23, 1965, four years before our first landing on the moon. He did not live to see space exploration shatter the illusions of his numerous followers. We now know with certainty that there is no life on the moon, and that Venus and the other planets in our solar system are unlikely to support intelligent life. Adamski, then, was either fooling us, being fooled by someone else, or having a psychological experience that he thought was real. He has also left us with at least one puzzle that keeps him from being entirely discredited.

In his account of his first alleged space flight he describes the experience in these words:

"Firkon motioned me to come to one of the portholes as he said, 'Perhaps you would like to see what space really looks like.'

"I soon forgot my disappointment as I looked out. I was amazed to see that the background of space is totally dark. Yet there were manifestations taking place all around us, as though billions upon billions of fireflies were flickering everywhere, moving in all directions, as fireflies do. However, these were of many colors, a gigantic celestial fireworks display that was beautiful to the point of being awesome."

Compare Adamski's words with those of astronaut John

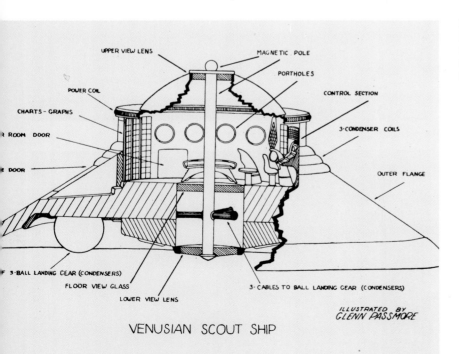

VENUSIAN SCOUT SHIP

UPPER VIEW LENS   MAGNETIC POLE

PORTHOLES

POWER COIL   CONTROL SECTION

CHARTS - GRAPHS

R ROOM DOOR   3- CONDENSER COILS

E DOOR   OUTER FLANGE

F 3-BALL LANDING GEAR (CONDENSERS)

FLOOR VIEW GLASS   3- CABLES TO BALL LANDING GEAR (CONDENSERS)

LOWER VIEW LENS

ILLUSTRATED BY
GLENN PASSMORE

Left: a diagram of a Venusian scout ship, based on Adamski's description after he had taken a trip with three extraterrestrials. One of them was his old acquaintance from the first contact.

A sequence of photographs taken by
Adamski on March 5, 1951.
Right: one of the Venusian scout ships has
been launched and a second one is just
emerging from the long cigar-shaped
mother ship.

Right: four of the scouts have appeared,
still hovering close to the mother ship, and
a fifth is being released from the bottom.

Right: six scout ships and the mother ship
in a cluster. All are enveloped in a bright
glow.

Glenn during the first manned orbital space flight:

"The biggest surprise of the flight occurred at dawn . . . When I glanced back through the window my initial reaction was that the spacecraft had tumbled and that I could see nothing but stars through the window. I realized, however, that I was still in the normal attitude. The spacecraft was surrounded by luminous particles. These particles were a light yellowish green color. It was as if the spacecraft were moving through a field of fireflies."

A coincidence? Adamski's account was first published in 1955. Glenn's flight was on February 20, 1962. Perhaps this can mean that Adamski had a genuine contact with extraterrestrials, but that his version of what happened was greatly embellished to make it acceptable to the general public.

Adamski never claimed that his experience was unique. The space people had told him that they were contacting others, he said. Naturally a whole crop of such stories soon appeared. Most offered nothing in the way of corroborative evidence, and each seemed to be vying with its predecessors to be more fantastic.

Serious UFO investigators, particularly those who believed the saucers were interplanetary, did not know which way to turn. Here was the evidence they had been waiting for, but now that it had arrived they did not want it. They believed that these stories of personal contact with space people would do more

Spacecraft From Venus

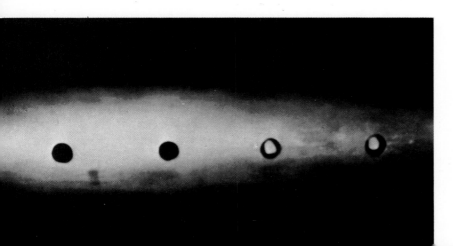

Left: a diagram of the mother ship showing the landing chute at the top and the launching chute at the bottom, as printed in the book *Flying Saucers Have Landed* by Adamski. On the left is the hangar deck where scout ships are stored when not in use.

Left: a photograph of Adamski (on the right) and a Venusian peering out of the portholes of a small mother ship. According to Adamski it was obligingly taken by another Venusian from a scout ship with Adamski's own Polaroid camera. Adamski made the common photographer's complaint that he could have used more film than he took with him.

Encounter with Strange Dwarves

Below: a copy of the sketch made by the painter and writer R. L. Johannis of two UFO pilots he encountered in the hills of Italy in August 1947. This was one of the earliest contact stories to get international news coverage.

Opposite: laughing dwarves from outer space. In 1954 they were said to have popped out of a spool-shaped flying saucer and accosted an Italian woman in the woods. They snatched one of her stockings and the carnations she was carrying. Then they disappeared.

harm to the cause of serious research than a thousand denials by military authorities. Some UFO enthusiasts abandoned their interest; others went back to studying obscure objects and lights in the skies while studiously ignoring claims of contact; but some persevered. Trying to establish a pattern in the contact accounts was practically impossible. It seemed that if every account were true, they must each concern visitations from different planets. There were giants, dwarfs, hairy creatures, smooth-skinned humans, and numerous permutations.

The two earliest contact stories to come to light happened in the year flying saucers became hot news, just 29 days after Kenneth Arnold made his famous sighting on June 24, 1947 and five years before Adamski's story. José C. Higgins, a survey worker, was with a group of others in an isolated area of Brazil when a large circular craft of grayish white metal descended from the sky with a whistling sound. Higgins' companions ran away. The huge UFO—it was about 150 feet wide—settled on the ground on curved metallic legs, and three figures got out. It was not clear whether they were male or female. They were about 7 feet tall, and their heads and bodies were covered with inflated transparent suits. Underneath they seemed to be wearing clothes made of brightly colored paper. They wore metal boxes on their backs, and all had identical heads—large, round, and with big eyes. They had no eyebrows or beards. Higgins found them strangely beautiful. They drew eight circles on the ground, indicating that the seventh was their home. This was later interpreted as meaning that they came from Uranus, the seventh planet. Higgins eluded their efforts to entice him aboard the spaceship, and watched for half an hour while they jumped and leaped around, tossing enormous stones. They then climbed back into their vehicle and vanished into the sky.

A month later in Italy, Professor R. L. Johannis, a well-known painter and writer, encountered a very different kind of space visitor. On August 14, 1947 he was out searching for fossils near a small village in the northeast of Italy. Equipped with a knapsack and a geologist's pick, he was making his way along a valley when he saw a large red lens-shaped object embedded in the mountain rock in front of him. From it protruded two antenna. Not far away he saw what he thought were two boys, but as he neared them he realized to his horror that they were strange dwarves. They were only about 3 feet high and their heads gave the impression of being larger than a normal person's. They wore dark blue overalls of a translucent material with vivid red collars and belts. Brown caps covered any hair they may have had, and their skin was green. Their mouths were small slits like those of fish, and their eyes were round and protruding like large yellow-green plums. As they came closer the professor found that he was rooted to the spot, petrified. Eventually he raised his arm, pointed at the spacecraft, and shouted to find out who they were and whether he could help them.

The dwarves seem to have mistaken his action as being hostile, and he was suddenly hit by a ray emitted from the belt of one of the creatures, knocked to the ground, and almost paralyzed. The dwarves came toward him, and one of them bent down to retrieve the pick he had dropped. Horrified, the professor saw

They Talked Till Dawn

While asleep in his truck at midnight on July 28, 1952, Truman Bethurum of Prescott, Arizona was awakened by strange visitors — short spacemen in coveralls. They took him to their spaceship and he talked to their beautiful female chief till dawn. Then he wrote this note, as printed in *Life* magazine in 1957: "To whom it may concern. If I am found dead it will be because my heart has stopped from the terrible excitement induced by seeing and going aboard a flying saucer."

Spaceship Inspection

One night in the summer of 1953 George Van Tassel lay sleeping in the desert near Giant Rock. To his surprise a spaceman woke him and offered to let him inspect a spaceship nearby. Van Tassel said yes and, clad only in shorts, was lifted straight up in an upright position. He had a good look at the spaceship, and was impressed by inventions that included a closet which cleaned clothes automatically with light. After a thorough inspection, Van Tassel left on a gravity nullifying beam.

It Rained Space People

Ruth May Weber of Yucca Valley, California had a psychic experience involving space people. She was told by a voice that Earth was already inhabited by space people, who would take over in case of a world disaster. This voice message was confirmed when she saw a shower of spacemen and women raining from the skies on Main Street in the town. They disappeared into the crowd entirely unnoticed by the ordinary Yucca Valley residents, who went on about their ordinary business.

that he had jointless green claws instead of hands. The two dwarves climbed back into the disk embedded in the rock, and in a few moments it shot out of the cleft. For a while it hovered over the terrified man. Then it shot away. It took Professor Johannis a long time to muster enough strength to get up from the ground. When he had partly recovered he noticed that his thermos was shattered, and that there was no trace of the metal container. An aluminum can and fork had disappeared, and so had his geologist's pick.

The year of 1952 in which Adamski claims first to have met space visitors was a vintage one for contacts. Mayor Oskar Linke reported that he and his daughter Gabrielle were four miles inside the Soviet zone of Germany on their way back to the Western zone when they came across two humanlike figures dressed in one-piece silvery suits. One had a flashing light on his chest. Nearby, resting in the forest glade, was a saucer, 50 feet wide and shaped like a large oval warming pan. When these beings heard the voices of Herr Linke and his daughter they immediately ran to their craft, which took off amid roars and flames. They watched as it whistled away over the treetops.

Next came Orfeo Angelucci whose alleged communications with the occupants of the saucers was on a telepathic level. He claimed to have seen a red light in the sky release two green disks. When these came to rest a few feet from his car a voice spoke in English. Then a screen appeared between the two UFOs, and the faces of a most noble man and woman appeared. His discourse lasted two hours. During another encounter he was taken aboard a saucer and given a ride.

Not all the reported contacts were so friendly. On September 12, 1952 a group of youngsters reported seeing a meteor land on a hill at Flatwoods, West Virginia. On the way to investigate they stopped at the home of Mrs. Kathleen Hill. She, her two sons, and a 17-year-old National Guardsman, Gene Lemon, decided to join the search. As they climbed the hill they saw a large globe which one said was "as big as a house." One boy heard a throbbing sound, another a hissing noise. The group's attention was suddenly drawn to a huge figure concealed in the branches of a tree. It was between 10 and 15 feet tall and had a blood-red face and glowing greenish orange eyes. When this monster floated toward them they fled hysterically down the hillside. Some of the group were violently ill for the rest of the night. The following day parallel skid marks were found on the grass, and a strange smell lingered over the ground.

This is one of the few contact stories involving a monster. Humanlike operators are far more common, though their appearance covers a wide range from the terrible dwarves seen by Professor Johannis to the beautiful female pilot whom Truman Bethurum, a mechanic, encountered in the Nevada desert. Her name was Captain Aura Rhanes, and she spoke English in rhyming couplets. She told him she came from a world behind the moon, the planet Clarion. Captain Rhanes was $4\frac{1}{2}$ feet tall, wore a black skirt, red blouse, and a beret. Her crew consisted of 32 men, all small like herself. After 11 meetings the space visitors were apparently frightened off by Bethurum's familiarity with them in a restaurant. He approached the captain

More and More Contact Stories

Below: one of the goblins from a UFO that besieged the home of the Sutton family of Hopkinsville, Kentucky, throughout one harrowing night.

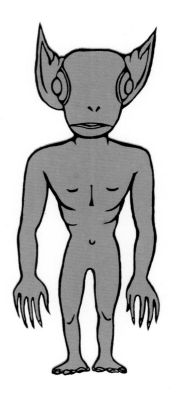

Right: Betty and Barney Hill. They sighted a UFO and subsequently discovered they had lost track of two hours and 35 miles of distance. During hypnosis to help Barney Hill overcome anxiety, the missing two hours were accounted for. The couple had been on a spacecraft whose crew examined them like laboratory animals, and released them with the promise that they would remember nothing.

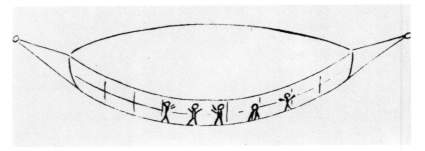

Right: Barney Hill's sketch of the UFO as it looked when they first saw it traveling parallel with their car on an empty road.

and eight of her crew in the restaurant saying: "Ma'am, we have met before? Say, ain't these the little guys from Clarion?" They all got up, walked to the exit, and vanished into thin air!

In February 1954 an English writer of thrillers, Cedric Allingham, claimed to have met a man from Mars at a lonely spot in Scotland. He said he had seen and photographed a flying saucer while out bird watching, and had continued walking along the coast after lunch. The saucer flew in from the sea, and he was able to photograph it as it came to rest 50 yards away. It was similar to Adamski's saucers with its three-ball landing equipment. A humanlike being with a very large forehead stepped out. A small tube was attached to his nose, presumably as an aid to breathing on Earth. He communicated by drawing in the writer's notebook. Topics touched on were the atom bomb, the Martian canals, and the shortage of water on Mars.

While those who claimed contact with space people were entertaining the public and making a good living out of it, there was a steady build-up of saucer activity that involved UFO landings and takeoffs without contact. Jacques Vallee, French-born astronomer, expert on computer techniques, and consultant to NASA on the Mars Map project, contributed to the book *The Humanoids* a report of 200 such sightings during 1954 alone. On November 8 of that year, for example, a man saw a light in a stadium in Monza, Italy and soon a crowd of 150 people gathered. They destroyed the barriers and rushed in to have a look. They saw a craft on three legs emitting a blinding white light. Figures in light colors and transparent helmets were seen standing close by. One had a dark face and a trunk or hose coming up to his face. They seemed to communicate with guttural sounds. The strange disk flew away without any noise.

A year later the Sutton family of Hopkinsville, Kentucky, were besieged by little monsters from a saucer. A visiting relative on his way home reported seeing a UFO land in a nearby field. Minutes later the family saw a small glowing figure approaching the house. It was no taller than 3½ feet, had a roundish head, elephantine ears, a slit mouth from ear to ear, and huge wide-set eyes. It appeared to have no neck, and its long arms ended in clawed hands. It dropped to all fours when it ran. Several of these creatures roamed around the outside of the Suttons' lonely home, climbing trees and clambering onto the roof. A gun was fired through a screen door at one of the figures. It was knocked over by the blast but ran away. After a night of terror, the family scrambled into a car and drove into town to report their experience.

In England in 1957, a Birmingham housewife claimed to have been visited by a spaceman in her own home. He did not arrive in a flying saucer, nor would the neighbors have seen him at the door. He simply materialized, accompanied by a whistling sound, in the living room of 27-year-old Mrs. Cynthia Appleton. He was tall and fair and wore a tight fitting plasticlike garment. He communicated by telepathy and produced TV-like pictures to illustrate his flying saucer and a larger master craft. He indicated to her that he came from a place of harmony and peace. At the

Kidnapped by a UFO Crew?

Below left: one of the occupants of the UFO that the Hills sighted. The information for the picture was given under hypnosis more than two years after the event. Until the deep trance therapeutic sessions, the Hills had no conscious memory of the experience.

Below: an artist's impression of the UFO crew members Betty and Barney Hill described as their abductors. They had two slits where nostrils would be, but no nose, and their mouth had no muscle. Their eyes were elongated and seemed to give a wider field of vision than we have. All wore dark clothing that looked alike.

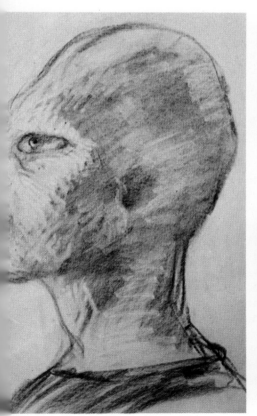

A Sexual Encounter!

Right: Antonio Villas Boas, a young Brazilian farmer, undergoing a medical examination after his strange encounter with the crew of a UFO. In 1957 he claimed to have had a passionate sexual experience with a beautiful tiny space woman, after he had had a blood sample taken from his chin. The doctor found that Villas Boas had been subjected to radiation, and that there were two unexplained needle marks on his chin.

Below: two drawings by Antonio Villas Boas of the UFO on which he had a trying experience. He was plowing a field alone when he was kidnaped by the crew.

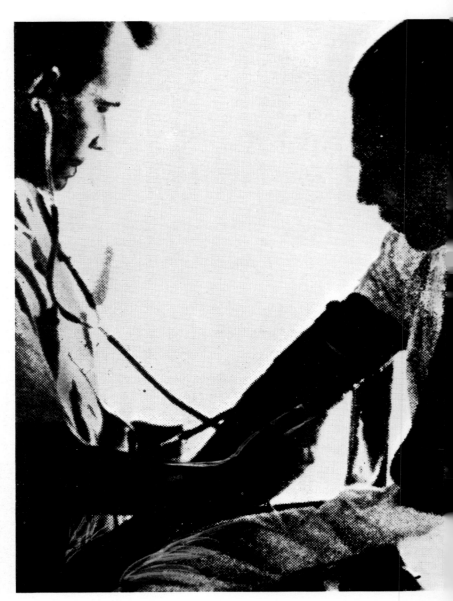

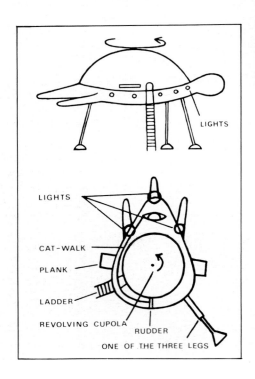

end of this and other visits he simply dematerialized.

The missing two hours in the lives of Betty and Barney Hill after a UFO sighting are regarded by many as the final proof that flying saucers are spaceships from other worlds. The couple was returning to their New Hampshire home on September 19, 1961 after a short vacation in Canada. During their journey they spotted a light in the sky that grew larger and appeared to be following a parallel course to their car. Then they saw a UFO a few hundred feet ahead of them. They stopped their car, and Barney viewed the huge structure through his binoculars. He saw two rows of windows from which figures peered out at him. He ran back to the car and drove off. They heard a beeping sound and felt drowsy. Two hours later they heard the noise again, and slowly became aware that they were driving 35 miles south of where they had seen the object. They could not account for their activities during the lost period. A compass reading showed that their car appeared to have been subjected to radiation.

They immediately reported their sighting to Pease Air Force Base and to the National Investigations Committee on Aerial Phenomena (NICAP), a UFO research group in Washington DC. It was not until three years later, when Hill was extremely sick and undergoing psychiatric treatment, that hypnosis revealed what seemed to have happened during the missing two hours. Both Betty and Barney Hill underwent hypnotic regression to uncover the experience, and their accounts agreed.

According to both of them, after hearing the odd beeping they were taken from their car by a strange group of men and carried onto a huge spacecraft. Separately they were laid on operating tables and carefully examined. He felt someone putting a cup over his groin. She had a large needle inserted into her navel. Samples of hair and toenails were taken. The spaceship's occupants were puzzled that his teeth came out whereas hers did not. The Hills were told that they would not remember their ordeal when released.

An even more bizarre abduction occurred in Brazil on October 15, 1957. Antonio Villas Boas, a 23-year-old farmer, had twice sighted bright unidentified flying objects. One day after the second sighting while plowing a field, he claimed that a huge egg-shaped luminous object appeared and landed in front of his tractor. Terrified, he leaped off and ran away only to be grabbed by the arms, lifted from the ground, and carried into the saucer by three figures in tight gray overalls and helmets. There were other occupants on board the saucer, and they all communicated with strange sounds resembling barks or yelps. Villas Boas was stripped naked and sponged all over. A blood sample was taken from his chin. Then he was left alone in a room with a couch. After a while a naked woman entered the room. She was about 4 foot 5 inches tall, and had the most beautiful body he had ever seen. Her hair was almost white, her eyes were large, blue, and slanted. She had high cheekbones, a straight nose, and a pointed chin. She moved toward him and embraced him. The result of this strange meeting between inhabitants of two worlds was a passionate sexual encounter. It was not all bliss, however. Villas Boas recounted that "some of the grunts that I heard coming from that woman's mouth at certain moments nearly spoiled everything, giving the disagreeable impression that I was with an animal."

The Villas Boas account is perhaps the ultimate in saucer contact stories. It may well have been discarded as the sexual fantasy of a young man were it not for the medical examination he underwent soon after reporting the encounter. Dr. Olvao T. Fontes reported that the farmer had been subjected to sufficient radiation to produce symptoms of radiation poisoning over a long period. At the point where blood was said to have been taken from his chin the doctor found two small hypochromatic patches. The skin looked smoother and thinner, as though it had been renewed.

If just a few of the contact stories are true, then we on Earth may be under close scrutiny by creatures from outer space. If the Villas Boas story is also genuine, then somewhere out in space a remarkable breeding experiment may now be in progress.

Yamski or Adamski?

E. A. Bryant, a retired prison officer who lived in southwest England near Dartmoor prison, was taking a walk on the evening of April 24, 1965. About 5:30 p.m. he arrived and stopped at an especially scenic spot. All at once he saw a flying saucer appear out of thin air about 40 yards away. It swung left and right like a clock pendulum before coming to rest and hovering above ground.

Although he was frightened, Bryant was curious enough to overcome his fears and stay to watch. An opening appeared in the side of the spaceship and three figures, dressed in what looked like diving suits, came to it. One beckoned to Bryant, and he approached the strange craft. As he did so, the occupants removed their headgear. He saw that two had fair hair, blue eyes, and exceptionally high foreheads. The third—who was smaller and darker—had ordinary earthly features.

The dark one talked to Bryant in fairly good English. Bryant understood him to say his name was "Yamski" or something like it, and that he wished "Des" or "Les" were there to see him because the latter would understand everything. He also said that he and the others were from the planet Venus. After the UFO took off, some metallic fragments were left on the ground near the place it had been. Later some small pieces of metal were indeed found there.

When Bryant reported his experience, investigators were struck by the fact that George Adamski, author of the best-selling book *Flying Saucers Have Landed*, had died the very day before. His collaborator on this book had been Desmond Leslie. Was there a connection between "Yamski" and Adamski, and "Des" or "Les" and Desmond Leslie?

Chapter 11
The Aliens Are Among Us!

Perhaps the most spine-chilling accounts of contacts with space visitors are those which claim that the men from outer space are living among us, in the form of humans. Are there spacemen, with glowing eyes, walking down our city streets, anonymous in the crush of the crowd? There is an ominous report of three mysterious men in black, who seem to appear again and again, each time attempting to censor news of UFOs. Who are these men in black? Is there an everyday explanation for their curious interest in our development in space? The evidence is strangely ambiguous.

As interest in flying saucers increased in the 1950s, some UFO enthusiasts began to seriously consider an extraordinary idea. Perhaps they should look for evidence of extraterrestrial visitors on the surface of this planet rather than in the skies. It might be that beings from outer space had already infiltrated our communities and were living among us, gathering information and making contact with those who were sympathetic to the possibilities of life in other worlds. This theory had already gained currency in the strange accounts of those who had claimed contact, called contactees in UFO jargon. Adamski and others, having first met the space people dressed in space clothes and close to their spaceships, later claimed to have met them dressed as we do, and frequenting hotels and restaurants. This belief made the need for supporting evidence in the way of photographs unnecessary. Anyone could claim to have met someone from space in their homes or on the streets.

One of the most famous contactees of this period was a New Jersey signpainter, Howard Menger, who had apparently been meeting space people from an early age. But it was in his army days during World War II, when a long-haired blond man in a car contacted him, that he realized their true identity. Some time later, a UFO landed near his parents' home in June 1946, and two men and a beautiful girl dressed in pastel "ski uniforms" stepped out. The space woman, who said she was 500 years old, told him to learn to use his mental powers so as to be ready for

Opposite: a science fiction comicbook cover. At first most of the speculation about the occupants of flying saucers ran along the lines of little green men. When the idea of space visitors that look just like us presented itself, the possibilities for contact were excitingly wide-open.

Right: is this woman a visitor from outer space? Her name is Marla Menger, and her husband Howard claims that she is an extraterrestrial being with whom he has been in contact since he was ten years old.

Far right: Howard Menger, the New Jersey signpainter whose alleged contacts with the world of Venusians, Saturnians, and other inhabitants of our solar system were wild and wonderful.

important events in the future. She said he should not talk about his experiences until 1957. It was in that year, during a rash of UFO sightings in High Bridge, New Jersey where he lived, that several witnesses claimed they had watched as Menger went out to speak with the space people. The full story was related in Menger's book *From Outer Space to You*. It was even more bizarre than Adamski's account, and told of his many contacts with space people posing as ordinary Earth people such as businessmen and real estate dealers.

During an early radio broadcast about his alleged contacts, Menger met a striking blonde girl named Marla who was one of a crowd that had gathered outside the studio. He later divorced his wife and married Marla who, he claimed, was from another planet. The next fascinating twist to the story came when Marla wrote a book entitled *My Saturnian Lover*. The Saturnian in question was Menger! His space visitors had told him that he was originally from Saturn. This was divulged during one of his many conversations with them, for they frequently dropped in for coffee at his home. During these visits he was also asked to cut their long golden tresses to make them look more like earthly inhabitants. He was taken for a flight to the moon from which he brought back some strange rocks. He said they were moon potatoes.

While many contactees painted a cosy picture of their encounters with the people of the UFOs, there was another more sinister aspect. This came to the public's attention in 1953 when the International Flying Saucer Bureau, a fast-growing saucer society in the United States, suddenly closed down. Albert K. Bender had started the organization in January 1952. It was nonprofit and soon attracted members and officers eager to solve the UFO mystery. But Bender found himself caught up in a series of events that terrified him. They culminated in his decision to close the IFSB less than two years later. In the last issue of the bureau's journal *Space Review*, the following statement appeared:

"The mystery of the flying saucers is no longer a mystery. The source is already known, but any information about this is being withheld by orders from a higher source. We would like to print the full story in *Space Review*, but because of the nature of the information we are sorry that we have been advised in the negative." The statement went on to tell the journal readers:

"We advise those engaged in saucer work to please be very cautious."

All that was known by most of Bender's UFO associates was that he claimed to have been visited by three men in black. Some believed he was perpetrating a gigantic hoax in order to gain personal publicity. Others believed he had hit on the right solution to the UFO puzzle and that the government, knowing this, had silenced him. He lost friends, and some of his closest saucer colleagues were openly hostile. Bender refused to reveal "the truth" until 1962. But when he told it in his book *Flying Saucers And The Three Men*, he was subjected to even greater derision. It read like the most incredible science fiction.

Bender claimed that soon after the Bureau opened he began to experience strange phenomena, including a telepathic phone warning that he should not investigate saucers any more. This was followed by the appearance of a glowing object in his room. The glow left a pungent smell of sulfur which was to become a regular feature of the manifestations. When at the movies once, he became aware of a man seated next to him with "two strange eyes, like little flashlight bulbs lighted up on a dark face." This figure disappeared, reappearing instantly in a seat on the other side of the saucer researcher. Thoroughly rattled, Bender called the manager; but by the time he had arrived with his flashlight the mysterious figure was nowhere to be seen.

Some months later, Bender went upstairs to his room after a

The Mystery of the Mengers

Left: Menger's trip to Venus. His first wife was a simple Earthling and he had to creep out of bed in the middle of the night to keep his rendezvous with spacecraft awaiting him in Field Location No. 2, on a farm. This artist's impression shows him viewing the wonders of Venus on a wide screen in a spacecraft that is passing over the planet. Menger's second wife Marla, who was said to be an extraterrestrial, confused things by claiming that he was not an Earth man either. She wrote about his true identity in her book *My Saturnian Lover*.

"We Are Among You"

Below: Menger and several other contactees from all over the earth, were given a Moon tour. This is one of the pictures he snapped as a souvenir of the trip, of which he wrote regretfully that he was not allowed to photograph any surface detail, people, or mechanical installations. In his picture, according to Menger's account, a Venusian is shown standing in front of his spacecraft after it has landed on the Moon's surface.

Opposite below: on the Moon. The round object is the Moon base, and a Venusian spacecraft is flying in the emptiness above it.

meeting, and noticed that someone had opened his door. A strange blue light emanated from within. As he entered, armed with a broom, he saw that the glow came from one corner. Some object seemed to be in the center of the light. Bender shouted, "Cut the kidding and come out of there!" and the glow began to fade, leaving the image of two glowing eyes for several minutes. Then the room was normal except for a strong smell of sulfur everywhere.

March 15, 1953, was the day chosen by the International Flying Saucer Bureau to send a telepathic message to occupants of flying saucers, asking them to make contact and help the world solve its problems. A time was appointed, and each member of the IFSB was asked to repeat mentally an identical message. Bender lay down on his bed to carry out his part of the program. As he repeated the message for the third time he felt the most agonizing pain in his head. He could smell a strange smell like burning sulfur, and small lights began to swim through his brain. He felt ice cold. As he opened his eyes he seemed to be floating above his bed. It was as if his soul had left his body and was looking down on it. Suddenly a voice seemed to come from in front of him. It was as though it permeated him without issuing an audible sound. He was warned not to delve any further into the mysteries of the Universe. He was told that "they" had a special assignment "and must not be disturbed by your people." When he tried to remonstrate the voice added: "We are among you and know your every move, so please be advised we are here on your Earth."

A few days later the mysterious communicators made an appearance to Bender. As with many other visits, their arrival caused the saucer researcher to suffer pain and discomfort. In his book he wrote:

"Blue lights appeared from nowhere and swirled about in the room. I grew dizzy as the areas above my eyes throbbed and again felt puffy. I stumbled to the bed and threw myself upon it. As I did so, I felt my body growing icy cold. I could feel I had quickly come under the complete power of someone or something.

"The room seemed to grow dark yet I could still see. I noted three shadowy figures in the room. They floated about a foot off the floor. My temples throbbed and my body grew light. I had the feeling of being washed clean. The three figures became clearer. They looked like clergymen, but wore hats similar to Homburg style. The faces were not clearly discernible, for the hats partly hid and shaded them. . . . The eyes of all three figures suddenly lit up like flashlight bulbs, and all these were focused upon me."

By telepathy the three visitors gave Bender a message. He was told that the UFOs had a purpose in visiting Earth, and that they must not be disturbed until their goal was accomplished. "As you see us here, we are not in our natural form. We have found it necessary to take on the look of your people while we are here." Bender was told that they had made numerous contacts with Earth with their spacecraft, and they had established a base at a remote spot on this planet. He had been selected as a contact because he was so ordinary. This meant no one would believe him if he betrayed their secret.

"We also found it necessary to carry off Earth people to use their bodies to disguise our own," Bender was told. They gave him a small disk, about the size of a coin. He was told to grasp it in his hand, turn on his radio, and repeat the word "Kazik" three times if he should wish to contact them.

At this point most serious UFO researchers reading Bender's account doubtless decided that their former colleague had taken

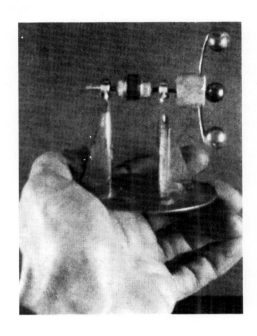

Above: Menger's "free energy motor," which he constructed one evening in his signpainting shop apparently under telepathic control. It was a kind of perpetual motion machine. The first and most successful version crashed through the roof and he was unable to duplicate it.

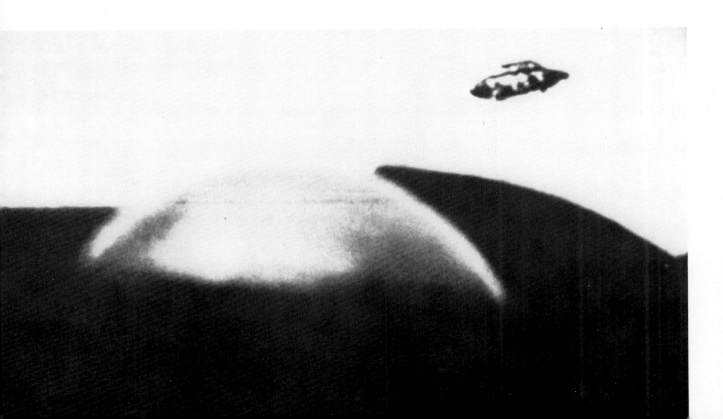

The Men in Black

Above: Albert K. Bender, a UFO investigator and founder of the International Flying Saucer Bureau. He became the center of curiosity and controversy when he abruptly closed the Bureau in 1953, saying only that he had done so after a visit from three mysterious men dressed in black. He refused to say anything more at all about it for seven years.

leave of his senses. But the one-time head of the civilian saucer investigation group had even more incredible material to relate. During his next contact with the three men in black he was somehow transported by them to a huge circular room with a glass dome, bathed in a brilliant glow. It may have been the interior of a flying saucer. He was greeted by a man who revealed that the purpose of the UFOs was to take a valuable chemical from our seas. To carry out their purpose, they had their people placed in strategic places everywhere including the Pentagon. They could destroy the world with a flick of a switch. Bender was shown various scenes on a three-dimensional screen, including one of a hideous monster. It was, he wrote, "more horrifying than any I have ever seen depicted in the work of science fiction or fantasy artists. The monster was alive." Without Bender noticing it, the speaker disappeared. Suddenly his voice came from the image of the monster, and he told Bender that that was his normal appearance, and they all looked like this on their home planet although they, of course, found nothing strange in their own appearance. He was told that there were three sexes on his planet: male, female, and a third of neither sex who were the exalted ones—the leaders. The females laid eggs that were stored and hatched as necessary. The planet from which they had traveled was many many light years away.

During another contact with this strange space race, Bender claims to have been taken to see their base of operation in the Antarctic, to have examined a flying saucer at close quarters, to have seen a flying saucer garage, and finally to have come face to face with an exalted one—a 9 foot bisexual with the same shining eyes as the others. This creature told Bender they had been visiting Earth since 1945 and would stay for about 15 years. "We have carried off many of your people to our own planet for means of experimentation, and also to place some of them on exhibit for our own people to see," this base commander told the saucer researcher. He warned that Bender would be taken if he got in their way. Back on Earth, Bender's eerie adventure continued. An invisible entity, whom Bender spotted only by the indentations on an armchair, visited his room. When the invisible form materialized Bender experienced "the worst fright of my entire life." He saw a 10-foot tall creature, greenish all over except for a glowing red face and glowing eyes. Bender fainted.

The space people's intrusions into Bender's life were not without compensations. On another visit to the bright glass dome, he was stripped by three beautiful women dressed in tight white uniforms. He was placed on an operating table and the three woman applied a liquid and "massaged every part of my body without exception, turning me over on my stomach and my sides. They expressed no emotion, neither that of revulsion nor enjoyment, as they carried out the matter."

An object was lowered over his body and he was bathed in a light that changed from lavender to deep purple. Afterward he was told that this was to ensure that he would enjoy good health for years to come. At the same time he was subjected to a ray that created an impulse in his body which their instruments could detect at all times. "We have found it necessary to do this in order to keep you under our constant surveillance. If at any time you reveal our secret we need only press a button in our laboratory and your body will be destroyed. It will disintegrate and very little will be left of it." That was the penalty for divulging his knowledge.

With that threat over him, and knowing "the secret of the saucers," Bender decided to wind up his organization. He was told that he was to keep the strange disk hidden. When it disappeared, he would know that the space visitors had accomplished their mission, and he could then discuss his contacts with them. According to Bender, the departure took place in 1960. He went to his strongbox where he kept the disk, having smelled the strange odor of sulfur. He threw open the lid and saw that it had disappeared. What is more, all his other possessions in that box had turned to dust.

He decided to write about the encounters although he knew, as he states in the book, that almost no one would believe him. But he claimed that "since the metal disk vanished and the visitors from space left our planet, flying saucer reports have decreased."

There is certainly no more reason for accepting Bender's incredible tale than that of other contactees. But there is one factor that has caused many investigators to ponder a little before dismissing it as horrific science fiction. That is the role of the three men in black. These mysterious men have cropped up time and again in the flying saucer riddle. In the early days it was assumed that they were government or CIA personnel, either taking a keen interest in the activities of certain UFO researchers or groups, or actively attempting to censor news of UFOs. Time and again the three men in black were reported to have called on UFO witnesses or researchers. They arrived in shiny black Cadillacs and were invariably described as small dark-skinned Orientals. They carried official looking cards or passes, and appeared to have a great knowledge of UFO phenomena. It was not until the late 1950s that some of the flying saucer authors began checking on the three men in black. They discovered that if they gave their identities, these proved to be false. Occasionally someone tried to follow one of the Cadillacs, only to lose it, often in circumstances which indicated that it had vanished or dematerialized. Though they usually turned up in threes, the men in black occasionally worked alone or in pairs. Typical of these reports is one cited by the Colorado Report, which had been carried out by the University of Colorado with the backing of the United States Air Force.

Case Number 52, occupying 18 pages of the massive report, deals with a series of UFO photographs taken by Rex Heflin, a California highway inspector. He sighted a metallic disk in the sky above Santa Ana Freeway and took three Polaroid photographs of it and one of a smoke ring the UFO is alleged to have left behind when it flew away. The case received widespread publicity, and he was interviewed by civilian UFO investigators as well as Air Force personnel. The El Toro Marine Air Station took a special interest in the case. Among his many callers Heflin received a man claiming to be from the North American Air Defense Command (NORAD), and he gave him the originals

"Exalted One from Outer Space

UFO enthusiasts were startled when the International Flying Saucer Bureau stopped all activity in late 1953. Although less than two years old, the organization founded by Albert K. Bender in Bridgeport, Connecticut had been prospering. Why did it shut down?

The only person who could answer this question was Bender himself, and he was not talking. Not until seven years later did he talk—and his tale was one of contact with an "Exalted One" from space.

Bender claimed to have had a direct interview with the Exalted One, who warned him of instant death if he continued to delve into the mystery of UFOs. That's why he disbanded the IFSB. But he learned many interesting things during his interview, and he revealed all when he got the sign he could.

In his interview, for example, Bender asked the space being if his people believed in God. The reply was that they did not, because they did not have the desire to "worship something" as Earth people do.

Another topic Bender discussed with the space visitor was whether other planets were inhabited. The Exalted One said that there had once been life on Mars, but that it was destroyed by invaders. The Martians had built beautiful cities and developed a vast system of waterways, but had not been as technologically advanced at the time of their end as we now are. Venus was developing life, the being said.

One of the most interesting exchanges between Bender and the Exalted One was about the Moon. Bender asked if the Earth people would ever reach the Moon, and he was told yes.

Seven years after Bender wrote this, men were indeed walking on the Moon.

of the photographs. These were never returned and NORAD subsequently denied that any of its representatives had visited the UFO witness. The sighting was on August 3, 1965. Fortunately Heflin had copied the original photographs, but the mystery of his sighting and the visitor remained unsolved. Strangely, when the Colorado team began investigating the Heflin case as part of its independent assessment of the flying saucer situation, the highway inspector received another mystery caller. This is how the Colorado report deals with the incident:

"According to the witness, on 11 October 1967, during the period when our investigation was beginning, an officer in Air Force uniform came to the witness's home in the evening and presented his credentials. Mindful of past experience, the witness studied them carefully. They gave the name Capt. C. H. Edmonds, of Space Systems Division, Systems Command. . . . The man allegedly asked a number of questions, including 'Are you going to try to get the originals back?' The witness claims that the man appeared visibly relieved when the witness replied 'No.' The officer also assertedly asked what the witness knew about the 'Bermuda Triangle' (an area where a number of ships and aircraft have been lost since 1800s).

"This alleged encounter took place at dusk on the front porch. During the questioning, the witness says he noted a car parked in the streets with indistinct lettering on the front door. In the back seat could be seen a figure and a violet (not blue) glow, which the witness attributed to instrument dials. He believed he was being photographed and recorded."

When the Colorado investigators followed up this lead they discovered that of the four "C. H. Edmonds" on the Air Force's list of officers, there was none with the corresponding rank and spelling. "All were of rather high rank and none should have had any connection with the Santa Ana case."

Afterward Heflin told investigators he believed his phone was being tapped and his mail intercepted. He also said that on three or four occasions his neighbors had told him that men in military uniform had come to his door during the day when he was not there.

Another case involving apparent imposters in uniform was investigated by writer John Keel in his book *Operation Trojan Horse*. In November 1966 two women had been watching flashing lights in the night sky in Owatonna, Minnesota, when one of the lights moved rapidly toward them, and then hovered a few feet above the ground at the far end of the field in which they were standing. One of the witnesses fell to her knees in a trancelike state. The other, Mrs. Ralph Butler, was astonished to hear a strange metallic-sounding voice emanating from her friend's lips. "What . . . is . . . your . . . time . . . cycle?" it asked. Mrs. Butler tried to give an answer explaining how the year was divided into days and hours and so on, and after several more questions the friend suddenly returned to normal and the object shot away. They were excited about the experience, but discovered that when they tried to discuss it with others they were both plagued with blinding headaches.

After reading one of Keel's UFO articles in a magazine Mrs.

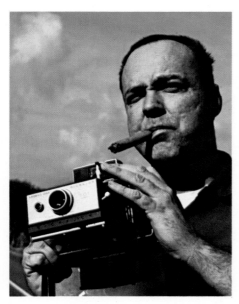

Above: Rex Heflin, who managed to get four Polaroid pictures of a UFO he sighted in 1965. All four originals disappeared after he gave them to some mysterious callers who claimed that they were official investigators, but who were unknown to the agencies they said they represented.

Below: William Hartmann, professor and investigator on the Condon Committee, working on the Heflin saucer sighting.

Butler wrote to him about their experience, and he phoned her. During an hour-long discussion she suddenly asked: "Has anyone ever reported receiving visits from peculiar Air Force officers?" The journalist said he had heard a few such stories, and she went on to give the following account:

"Well, last May (1967) a man came by here. He said he was Major Richard French, and he was interested in CB [citizen's band radio] and in UFOs. He was about 5 feet 9 inches tall with a kind of olive complexion and pointed face. His hair was dark and very long— too long for an Air Force officer, we thought. He spoke perfect English. He was well educated."

He wore a gray suit, white shirt, and black tie, and everything he was wearing was brand-new, she said. He drove a white Mustang whose license number her husband noted, and later found out it was a rented car from Minneapolis. The man said his stomach was bothering him and she advised him to eat jello.

Imposters in Air Force Uniforms

Below: the photographs that Heflin took through his truck window. The fourth shows the smoke ring that the UFO left as it vanished. Hartmann discovered that by hanging a Leica lens cap just a few inches outside the window, a picture very similar to the second photo could be achieved.

The Evidence Accumulates

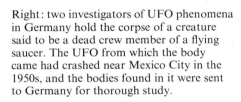

Right: two investigators of UFO phenomena in Germany hold the corpse of a creature said to be a dead crew member of a flying saucer. The UFO from which the body came had crashed near Mexico City in the 1950s, and the bodies found in it were sent to Germany for thorough study.

She said that if it continued to bother him he could come back for some, which he did the following morning. Mrs. Butler sat him at her kitchen table and produced a big bowl of jello. To her surprise he picked the bowl up and tried to drink it. "I had to show him how to eat it with a spoon," she told Keel. Later the same man turned up in Forest City, Iowa, and visited some of the Butlers' close friends.

Research has shown that there was a Richard French in the Air Force in Minnesota, but he did not even remotely answer to the description of the mysterious "major" who called on Mrs. Butler.

It has not escaped the attention of some UFO researchers that many of the humanoid figures reported near landed flying saucers are of similar build to the men in black and have the same facial characteristics. Their skin is described as olive-colored or tanned, their eyes are said to be large but with an Eastern cast or slanted, and their faces are sometimes described as pointed. As far back as the 1897 aircraft sightings in the United States, a certain Judge Lawrence A. Byrne came across a strange looking object anchored to the ground. "It was manned by three men who spoke a foreign language," he told the *Daily Texarkanian*, "but judging from their looks, would take them to be Japs. They saw my astonishment and beckoned me to follow them, and on complying I was shown through the ship."

Not having been able to solve the saucer riddle, some of the researchers have concentrated their studies—and their imagina-

Far left: the spirit photography of the 19th century takes a modern swing in the 20th— a lawyer from the west of England took a picture of his young daughter and was astonished to discover the image of a "spaceman" behind her when the photograph was developed.

Far left: the spirit photography of the 19th century takes a modern swing in the 20th— a lawyer from the west of England took a picture of his young daughter and was astonished to discover the image of a "spaceman" behind her when the photograph was developed.

Left: Lonnie Zamora, a New Mexico patrolman who chased a speeder and was distracted by a descending UFO throwing out a flame. He found it landed in a gully, with two aliens in light clothing. He took cover, and the craft took off again. When another policeman came along to help Zamora, who was agitated, he verified the presence of landing marks and smoldering plants where the UFO had set down.

tions—on the pursuit of the men in black. They have come up with some surprising conclusions. The men in black appear to have a mission, not always sinister. Their role might appear to be one of guiding the human race, or at least endeavoring to influence the course of its history. Some claim that they tried and succeeded when they brought to the attention of King Herod's court the imminent birth of Christ. The three wise men with Oriental features and dark skin who "came from the East" were the men in black—or the biblical equivalent. Not only did they show the same preknowledge of events that their modern equivalents demonstrate, but they even led the way to Bethlehem. They never returned to King Herod, as they had promised, but "went home by another way."

The Bible is almost as rich in stories of the three men— usually regarded by scholars as angels—as it is in alleged UFO accounts. In Genesis there is a description of three men who appeared before Abraham and ate and drank with him.

Other Bible accounts bring to mind once more the frightening encounters that saucer researcher Albert Bender claimed to have had with the men with shining eyes. Enoch, we are told, dreamed of two very tall men "such as I have never seen on earth. And their faces shone like the sun, and their eyes were like burning lamps. . . . They stood at the head of my bed and called me by my name. I awoke from my sleep and saw clearly these men standing in front of me." This might be regarded as just a coincidence, but the story continues with a familiar theme. The men took him

More Encounters

into the sky on a tour of "seven heavens." On his return to Earth he wrote an account of his experiences which is *The Book of the Secrets of Enoch*.

Supporting evidence is to be found in The Book of Daniel. The famous prophet saw "wheels as burning fire" and a figure that came down from a "throne" in the sky. The being's hair was like pure wool. It was dressed in a white robe with a gold belt, and it had a luminous face and two glowing eyes.

"And I Daniel alone saw the vision: For the men that were with me saw not the vision; but a great quaking fell upon them so that they fled to hide themselves.

"Yet I heard the voice of his words; and when I heard the voice of his words, then was I in a deep sleep on my face, and my face toward the ground."

Some might argue that Bender had read these biblical stories and based his own shining-eyed visitors on the descriptions of Enoch and Daniel. But that does not explain the many similar creatures seen by UFO witnesses.

On October 9, 1954, for example, a French farmer riding his bike near Vienne, France, saw a figure dressed in a diving suit. It had boots without heels "and *very bright eyes*." Two headlights were placed one above the other on its very hairy chest. It disappeared into the forest. Nine days later in Fontenay-Torcy, France, a man and his wife saw a red-colored cigar-shaped object in the sky. It dived toward them and landed behind some bushes. Suddenly the couple were confronted by a bulky individual just over 3 feet high whose eyes *gleamed* with an orange light. One of the witnesses lost consciousness.

On the beach at Ain-el-Turck near Oran, Algeria, a small man *with glowing eyes* was seen six days later. The following month, on November 28, two truck drivers in Venezuela found the road blocked between Caracas and Petare by a luminous sphere about eight to 10 feet wide, hovering about six feet above the ground. One got out of the truck and was approached by a small bristly dwarflike creature with claws and *glowing eyes*. A tussle took place and the driver was thrown 15 feet. He drew a knife to protect himself but found that stabs had no effect on the dwarf. The other driver ran for the police. Two similar creatures appeared from the bushes, apparently carrying stones and earth samples. They rushed into the sphere, and one of them blinded the embattled driver with a light to enable the third creature to flee. The sphere took off.

These incidents, all occurring in the space of a few months, do not prove Bender's story, of course; but they do illustrate that glowing eyes are a feature of some UFO accounts. We have also seen that similar beings appear to have been encountered during biblical times. The really sinister aspect of Bender's story is his assertion that these space visitors can take the form of humans, or even possess the bodies of abducted humans, in order to live on this planet. In fact, almost every contactee has claimed that the visitors from outer space, whatever their real guise, are living ordinary lives on Earth as part of an infiltration plan.

If the space people's purpose is simply to prepare the human race for ultimate contact with a superior space civilization, we

Below: Joe Simonton with one of the peculiar cookies he got from the UFO crew. Analysis showed that they were made of corn and wheat flours, but the origin of that type of wheat was unknown.

have nothing to fear. But some UFO writers are worried. They point out that the men in black have mysteriously appeared in cases that can have nothing to do with such an ultimate goal. Amateur sleuths investigating President John Kennedy's assassination, for example, have come across these enigmatic individuals. The American writer John Keel comments in his book *Our Haunted Planet*: "Here the black Cadillacs and the slight, dark men in black suits are viewed as Cubans and CIA agents. Paranoia runs high because now over 50 witnesses, reporters, and assassination investigators have met with sudden death, some under the most suspicious circumstances. The full story of Kennedy's murder in Dallas in 1963 is filled with incredible details, many of them similar to things found in the most mysterious of the UFO incidents. Photos and physical evidence have vanished or been tampered with just as in so many UFO cases."

Probably there is no link. Probably the men in black had no involvement in the assassination of the president. But strange coincidences bedevil the history of flying saucer research, leading investigators along new and bizarre paths and producing more and more fantastic theories to solve the puzzle.

It is probably also just a coincidence that Dallas, Texas—scene of Kennedy's assassination—has been a favorite haunt of flying saucers since the early days of the UFO era.

Below: an artist's impression of the UFO that reportedly landed in the yard of Joe Simonton's Wisconsin chicken farm in 1961. In it were three small dark men, one of whom asked Simonton for water. As he gave them a pitcher of water he saw that they seemed to be cooking, and asked for a sample of the foodstuff. It appeared to be a perforated cookie. The man gave him four, and then the UFO whizzed out of sight.

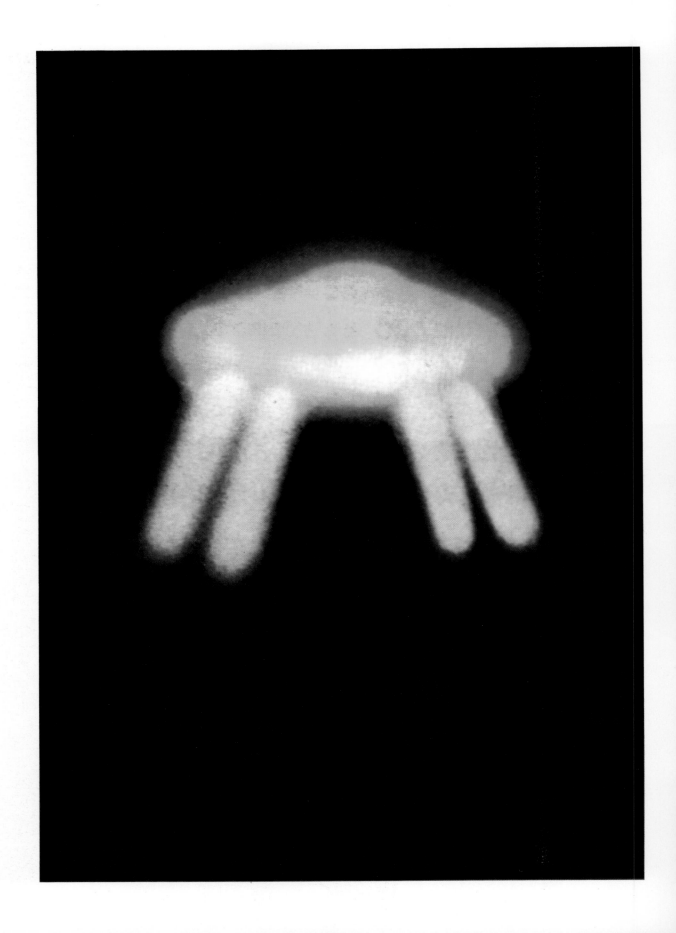

Chapter 12 UFOs Today

Well-publicized UFO sightings have been with us now for 30 years. What conclusions can be drawn about their validity and reliability? What is the quality of evidence of the more recent sightings? Since official government agencies have closed their investigations, how are UFO reports being checked into and collected? And even if specific sightings appear to be genuine, are we any closer to an explanation of what it was that the eyewitness actually saw? However many reports are dismissed, there always seem to be new observations of strange objects in the sky. Is there any way to solve the mystery?

According to a Gallup poll of 1966, more than five million people in the United States claimed that they had actually seen a flying saucer, and about 50 million—nearly half the adult population—believed that unidentified flying objects were real. But the official attitude continued to be negative. In 1969 the Condon Committee, which with the backing of the Air Force had been studying the problem for several years, finally published its massive report. This concluded firmly that UFOs did not exist. On the basis of its recommendations the Air Force closed down Project Blue Book, its own official inquiry into flying saucers. But the public, as ever, remained unimpressed by official pronouncements. Another Gallup poll taken in 1973 revealed not only that half the adult population in the United States still believed in UFOs, but also that by then more than 15 million claimed to have seen one, an increase of 200 percent in seven years.

Since the military has washed its hands of UFO investigations, the long-established civilian organizations like the National Investigations Committee on Aerial Phenomena (NICAP) and the Aerial Phenomena Research Organization (APRO) are steadily gaining in stature. Many scientists and other highly qualified investigators belong to these and other research groups both in the United States and other countries. Well-known figures in the world of astronomy—for example, Dr. J. Allen Hynek and Dr. Jacques Vallee—are conducting their own research projects. Their books have added a respect-

Opposite: this brightly colored UFO was sighted and photographed over France at about midnight on March 23, 1974. Still, today, such sightings continue to be reported, many by people who are just as baffled and stubbornly insistent that they saw some kind of flying object as were the first UFO-sighters some 20 or so years ago.

More Scientific Study of UFOs

able image to the field of UFO literature, which has seen too many sensational publications for its own good. Their work is concerned with quality of sightings rather than quantity, and they look for patterns in the appearance and behavior of UFOs that will help them build up recognized types of objects. Once types have been established, they might be able to conjecture further about the composition and source of power of flying saucers.

Typical of the sightings that interest these researchers were the remarkable events around Levelland, Texas, on the night the Soviet Union launched its second satellite carrying a dog into orbit on November 2, 1957. It was at 11 p.m. that the police received the first report. A truck driver and his companion had encountered a brilliantly lit torpedo-shaped object about 200 feet long. It had passed low over their truck at a rapid speed, apparently causing the lights to go out and the engine to stop. The men were petrified. As the object moved away, the truck lights came on again and the engine started.

An hour later the police received a call from a man who had come across a brilliant egg-shaped object about 200 feet long. It was sitting in the road about four miles east of Levelland. As he approached his engine failed and his headlights went out. He was about to get out of his car when the object rose and disappeared. He then found that his engine and lights were working again. The next report was from a man who had also seen a bright object sitting in the road 11 miles north of Levelland. His engine also stopped and his lights went out, returning to normal as soon as the object flew away.

A freshman from Texas Technical College about nine miles east of the town had trouble with his car at about this time. The engine stopped and the lights went out. It was not until after he checked under the hood that he noticed an oval-shaped object, flat on the bottom, sitting on the road ahead. He estimated its length at about 125 feet, and said it appeared to be made of a metallic or aluminumlike material. It glowed blue-green. He was scared, but, unable to start his car he sat and watched it for several minutes. It soon rose vertically into the sky and disappeared "in a split instant."

As further reports came in the police decided to take a look for themselves. Though none experienced close encounters or interference with their cars, two policemen reported seeing bright lights for just a few seconds. Levelland Fire Marshall Ray Jones was in the same area that night, and reported that his car headlights dimmed and his engine sputtered at the very moment he spotted "a streak of light" in the sky.

A total of 15 independent calls were received by the police that night. Every caller was excited. None of those who got close to the strange illuminated craft had any doubt about its existence. No satisfactory explanation has been offered for the Levelland episode, and since then the same interference with car engines, headlights, and radios has been reported on numerous other occasions.

The Levelland case is given in detail in *The UFO Experience* by J. Allen Hynek, former astronomical consultant to the Air Force's Project Blue Book. Dr. Hynek is an eminent astrono-

Below: Dr J. Allen Hynek, an astronomical consultant to Project Blue Book, remains convinced that many sightings were explained away too lightly. He believes that a methodical correlation of the various sightings—however bizarre some seem— will help establish recognized types of sightings and show up any patterns that exist.

mer. He is director of the Lindheimer Astronomical Research Center at Northwestern University and Chairman of the Northwestern Astronomy Department. He was for several years an associate director of the Smithsonian Astrophysical Observatory at Cambridge, Massachusetts, and headed its NASA-sponsored satellite tracking program. Since the early flying saucer days he had been employed as a scientific consultant on UFOs by the Air Force. As such he had been responsible for explaining many allegedly strange sightings as perfectly normal astronomical phenomena. But he also soon found that, though many sightings could be accounted for by misidentification of astronomical or other phenomena, there still remained some for which there was no easy explanation, and which were puzzling even to experts like himself. He became increasingly disenchanted with Project Blue Book's methods of working. The team was undermanned. Records were kept in chronological order with almost no cross-references. Worst of all, the investigators on the whole seemed more concerned with slapping a quick explanation on each sighting, rather than attempting to evaluate the problem in depth. The last straw for Hynek and those like him was the imbalanced Condon report that effectively caused the Air Force to abandon its investigation into UFOs.

Hynek likens the present situation on the UFO problem to that surrounding the meteorite debate nearly two centuries ago when meteorites were dismissed as "old wives' tales." Their existence was denied by science, and people who reported them were ridiculed and regarded either as mad, deluded, or hoaxers. We know now that those who saw meteorites were witnessing a real physical phenomenon. Stones did and do drop out of the sky, but Hynek does not know of any astronomer who had one land at his feet. In the same way, the evidence for UFOs depends largely on the accounts of eyewitnesses without scientific backgrounds, and there are still those scientists who adopt the attitude: "I won't believe it until I see it myself."

The problem is that UFO researchers can be just as biased in the other direction. Many adopt a theory about the flying saucers and then look for evidence to support it. Since UFO phenomena are so varied, it is not difficult to find several sightings to demonstrate the truth of almost any theory. Such researchers carefully omit other sightings which do not fit their pet ideas. Investigators such as Dr. Hynek and Jacques Vallee, the well-known astrophysicist and mathematician, insist that all aspects of the strange phenomenon must be studied if we are to solve the UFO puzzle. Their books deal with analysis of types of UFOs—such as nocturnal lights and daylight disks— as well as types of encounters ranging from close proximity aerial sightings to landings of craft and crew. There are recognizable patterns into which UFOs fall, but there are many contradictions that cause continuing confusion in UFO circles. So while the scientists search for tangible physical proof, others have turned to the occult in an attempt to find explanations for many of the astonishing incidents surrounding UFO cases.

Typical of the incidents which have moved thinking toward the occult was Patrolman Herbert Schirmer's encounter with a

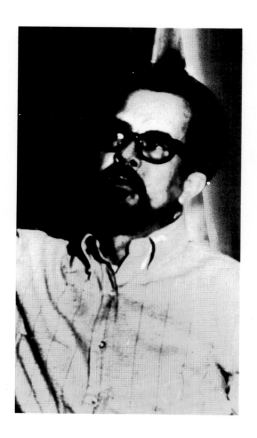

Above: Jim Lorenzen, a civilian deeply interested in methodical UFO research. In 1952 he and his wife Coral, who is also heavily involved in collecting and collating UFO reports, founded the Aerial Phenomena Research Organization. It acts as a clearing house for reports.

Their Experience Was "Very Real"

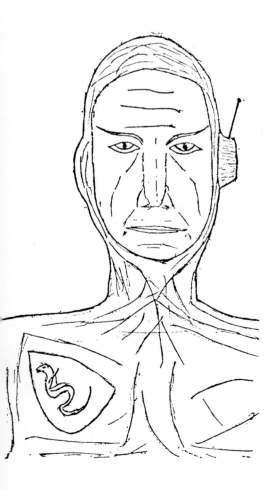

Below: Schirmer's drawing of the crew leader of the spacecraft.

UFO in Ashland, Nebraska in 1967. It was 2:30 a.m. on the morning of December 3. As 22-year-old Officer Schirmer was approaching an intersection on Highway 63, he saw an object ahead with a row of flickering lights. When he put on his main headlights, however, the object shot into the sky and disappeared. Half an hour later, back at the police station, his log book entry contained these words: "Saw a flying saucer at the junction of highways 6 and 63. Believe it or not."

Schirmer went home with a headache and a buzzing noise in his head which prevented him from sleeping. He also noticed a red welt that ran down the nerve cord of his neck below the left ear. It was not until Officer Schirmer was investigated by the Colorado team that he discovered he had apparently been contacted by the occupants of the UFO.

During time regression hypnosis his subconscious filled in the details of 20 missing minutes in his police report, immediately after seeing the UFO. He described following the object after it had landed in a field. He tried calling the police at Wahoo, Nebraska, to report the incident but his radio, lights, and engine were not working. The object, like a bright football, had three telescopic legs that descended as it settled on the ground. Then beings stepped out of the UFO shooting a greenish gas at him and his car. One of them pointed a bright ray that paralyzed him and caused him to pass out.

His next hypnotic memory was of being grabbed by the neck by one of the beings who had humanlike features. He was asked if he would shoot at a space ship. When he said he would not, Schirmer was told he could come aboard for a few minutes. The occupants were between $4\frac{1}{2}$-feet to 5-feet tall, and wore close-fitting silvery-gray uniforms and an antenna on their hoods or helmets. Their heads were thin and longer than human, with a flat nose, a slit for a mouth, and slightly slanted eyes that did not blink. They had a gray-white complexion. Schirmer, while still under hypnosis, was able to describe the inside of the flying saucer and its equipment. The occupants also projected pictures on a screen. Though the leader spoke to him, he was told his mind was also receiving simultaneous data input. Schirmer learned that they had come from a nearby galaxy. They had bases on Venus and other planets including Earth. On Earth their bases were underground or under the ocean.

"They have no pattern for contacting people," said Schirmer during the trance interrogation. "It is by pure chance so the government cannot determine any patterns about them. There will be a lot more contacts . . . to a certain extent they want to puzzle people."

After about 15 minutes inside the UFO, Schirmer was told not to reveal that he had been on board the craft. He was simply to report the first stage of his sighting. The beings appeared to have planted this hypnotic suggestion in his mind, for he had no conscious memory of his extraordinary experience although he did feel troubled and uneasy. It was only under hypnotism that the details came out. In this his experience was like that of Betty and Barney Hill in 1961, who only recalled a visit on board a spaceship under hypnosis several years later.

Six years after the Schirmer incident in October 1973, two

shipyard workers encountered the occupants of a flying saucer in Pascagoula, Mississippi. It is one of the most celebrated contacts of the 1970s. Though they did not need hypnosis to remember what took place, they both separately underwent hypnotic-regression, and their stories under hypnosis tallied. They revealed the most amazing information.

Charlie Hickson and Calvin Parker were fishing on the west bank of the Pascagoula River at about 7 p.m. when they spotted a strange object, emitting a bluish haze, about two miles away. It came toward them to within about 30 yards, and hovered three or four feet above the river. Both men were petrified. Three occupants emerged and floated toward them. Parker, aged 19, got hysterical and passed out. Hickson, aged 42 and known for his calm and coolness, was rigid with fear. One of the creatures emitted a strange buzzing sound. The two others, who had pinchers instead of hands, lifted Hickson up under the arms and glided on board their craft with him. Their vehicle was oblong and about 8-feet high with an opening at one end. Outside it was lit by a strange blue light and inside it glowed brightly. The creatures placed Hickson in a horizontal position where he floated while they examined him with a large eyelike device. They left him alone for a while and then returned. He tried to talk to them but they took no notice of his questions. Hickson was too scared to get an accurate idea of what they looked like. They seemed to be about 5-feet tall, and to have something projecting where a human nose would have been. Beneath this projection was a slit, but it never seemed to move. The head apparently sat on the body without any neck. He had no remembrance of eyes or hair. When they finally let him go, the vehicle made a buzzing sound and disappeared.

The two men were interviewed by Sheriff Fred Diamond and Captain Glen Ryder at the Sheriff's office in Pascagoula four hours after their terrifying experience. Their testimony was tape recorded. Then, to see if the whole thing was a hoax, they were left alone and the tape recorder was left running without their knowledge. Hickson sounded shaky and Parker was frantic as he recalled how his arms could not move.

Parker: "I passed out. I expect I never passed out in my whole life."

Hickson: "I've never seen nothin' like that before in my life. You can't make people believe—"

Parker: "I don't want to keep sittin' here. I want to see a doctor—"

Hickson: "They better wake up and start believin' . . . they better start believin'. . . ."

Hickson (later): "They won't believe it. They gonna believe it one of these days. Might be too late. I knew all along they was people from other worlds up there. I knew all along. I never thought it would happen to me."

Parker: "You know yourself I don't drink."

Hickson: "I know that, son. When I get to the house I'm gonna get me another drink, make me sleep. Look, what we sittin' around for? I gotta go tell Blanche . . . what we waiting for?"

Parker (panicky): "I gotta go to the house. I'm gettin' sick.

The UFO Chase

One of the wildest UFO chases on record took place early in the morning of April 17, 1966. It started when Dale F. Spaur, deputy sheriff of Portage County, Ohio, stopped at a stalled car and spotted the brightly lit flying object that had been reported to his office. He was ordered to follow it. He and an assistant raced after it in their patrol car for about 70 miles, sometimes driving over 100 mph to keep it in their sight.

Forty miles east of the point at which he started, Spaur met Officer Wayne Huston of East Palestine after having talked to him by car radio. This officer saw the flying saucer too, and described it as "shaped something like an ice cream cone with a sort of partly melted top." The chase continued into neighboring Pennsylvania, ending in Conway. Officer Frank Panzanella of that town came to the scene when he saw the other policemen, and told them he had been watching the shining object of the chase for about 10 minutes. All four observers then saw the UFO rise straight up left of the Moon, and disappear.

The United States Air Force Project Blue Book investigated the Spaur-UFO chase, and labeled it as a sighting of Venus. The four independent observers involved do not believe that was the right conclusion.

CALVIN PARKER

CHARLES HICKSON

I gotta get out of here.''

Hickson left his fishing companion alone. When the investigators played back the tape they heard the young man talking to himself.

Parker: "It's hard to believe . . . Oh God, it's awful . . . I know there's a God up there. . . .''

He then prayed and his words became an inaudible whisper.

Journalist Ralph Blum gives a detailed account of the Hickson–Parker case in his book *Beyond Earth: Man's Contact With UFOs*. During his own investigation of the contact report he interviewed Sheriff Diamond and asked if he believed the statements made by the two men. "First thing they wanted to do was take a lie detector test," the sheriff replied. "Charlie—he was shook bad." He added that you don't see a man of Hickson's age "break down and cry from excitement unless it's something fierce happened."

Dr. Allen Hynek, who was present during the time-regression hypnosis that both men agreed to undergo, declared later:

"There's simply no question in my mind that these men have had a very real frightening experience, the physical nature of which I am not certain about . . . They are absolutely honest. They have had a fantastic experience and also I believe it should be taken in context with experiences that others have had elsewhere in this country and in the world."

Dr. James Harder, professor of Engineering at the University of California in Berkeley and consultant to one of the largest civilian UFO organizations, the Aerial Phenomena Research Organization, supervised the regression hypnosis with a psychiatrist. His opinion was that "the experience they underwent was indeed a real one. A very strong feeling of terror is practically impossible to fake under hypnosis."

In the case of Patrolman Schirmer and Betty and Barney Hill the UFO occupants appear to have been able not only to control the minds of Earth people but also to obliterate their experiences from their conscious minds.

"We have no way of knowing how many human beings throughout the world may have been processed in this manner, since they would have absolutely no memory of undergoing the experience, and so we have no way of determining who among us has strange and sinister 'programs' lying dormant in the dark corners of his mind."

This is the view of American author John Keel, a leader among those UFO investigators who believe that flying saucers are not a physical phenomenon, and do not consist of solid matter. He thinks that UFOs have superior powers and exist alongside our world but on a totally different time scale. They are usually invisible but are capable of materializing in our world. They can take on any shape and travel at incredibly rapid speeds. Keel calls the inhabitants of the UFOs ultra-terrestrials and speculates on the reasons for their appearances to man. He suggests that perhaps the Universe is controlled and permeated by one gigantic intelligence or energy force which is all-knowing, and which is outside our concept of time and space. Throughout the Universe there are other lesser intelligences which exist at different time frequencies. The higher among these can to some extent manipulate the lower, though all are subordinate to the one main force. If this supreme intelligence wished to communicate with a fairly low form of intelligence such as man, it might choose to do so through an intermediary intelligence such as the ultraterrestrials. The information would be too complicated to be given at once, so it would be given gradually in small pieces over a long period of time. The flaw in this argument seems to be that misleading and contradictory information is often the result of exchanges between humans and ultraterrestrials. Keel has two possible explanations for this. One is that misleading information is deliberately given whenever it is felt that we are not yet ready to accept the truth. The other is that the occupants of the UFOs may be deliberately lying for some purpose of their own.

Keel believes that the ultraterrestrials are the same entities responsible for psychic phenomena and Spiritualism—another area of investigation that is full of conflict and confusion. Just as false and mischievous spirits are reported at seances, so the UFO entities appear to be deliberately trying to confuse us, says Keel. They set out to give the impression that they are from outer space to throw us off the scent, and at the same time they deliberately change their appearance and stories so that investigators will not be able to discover their true origin. The UFOs' occupants have been described variously as tall, average, and small; green, gray, and dark-skinned; of normal appearance, fishlike and monsterlike. They are said to have come from practically every planet in our solar system as well as from others in the furthest reaches of the Universe.

Either every contactee has lied about his or her experiences or else we are being visited by many different space races. Most contactees are simple, sincere people who appear to be convinced that they have had a real experience. Keel has interviewed 200 contactees who have not sought publicity about their encounters—and he has this to say about them in his book *UFOs: Operation Trojan Horse*:

"They do not write books or go on lecture tours. They show little or no interest in UFO literature. Some of them begin to experience personality deterioration after their initial contact. Others find their previously normal lives disrupted by nightmares and peculiar hallucinations. Poltergeists (noisy, invisible ghosts) invade their homes. Their telephones and television sets run amok. My own educated guess is that there may be 50,000 or more silent contactees in the United States alone. And new ones are being added to the list every month." Many of them may not even know they have been contacted unless, like Schirmer and the Hills, they underwent regression hypnosis to try to find out why they felt anxiety.

Since the late 19th century, when people first began reporting conversations with beings who landed in aerial craft, the witnesses have not been believed. But Keel insists that these contactees have been telling us *what they were told* by the ufonauts," and, he insists, "*the ufonauts are the liars, not the contactees.*"

His theory began to evolve after a study of the great American UFO panic of 1896–7. The aerial craft that people reported

The Anglers' Experience

Opposite left: pictures from a television report on Calvin Parker and Charles Hickson, who claim to have encountered a UFO while fishing. They were taken aboard by strange faceless creatures who examined them. The man at the top is Calvin Parker; the older man below is Charles Hickson. The drawing is their version of one of the creatures.

Below: Dr Allen Hynek as he questions and interviews Parker and Hickson.

Below: whatever else may be said about the creatures who reportedly visit our Earth, they are certainly varied. This is an artist's accurate reconstruction of one of three seven-foot tall entities sighted by Jose C. Higgins in Brazil in July 1947. They wore transparent suits inflated like rubber bags with metal boxes on their backs. They had large round bald heads, huge round eyes, no beards, and immensely long legs. They tried to capture Higgins, who escaped by hiding from them.

seeing looked much like the airships then being built and designed, but not yet flown in the United States. They were often seen on the ground and their occupants, who were usually described as foreigners, appeared to be repairing them. They frequently gave accounts of where they had come from and where they were heading, together with descriptions of how their airships worked. The simple farmers who had these encounters believed they were humans with advanced airships—about which they had read much in their newspapers. But Keel's examination of these reports and an investigation of some strange side issues reveals that no one on earth could have produced the craft, nor could they have been powered in the way the occupants claimed. Moreover, if these really were visitors from space, what were they doing cruising around in crude and cumbersome airships that needed so many running repairs? Clearly they had not traveled vast distances in space in such craft. Keel decided that the ultraterrestrials were responsible, and that it was no more than an elaborate smokescreen to conceal their real purpose and identity. In the 20th century, they confused us again by taking on a modern image in line with fast-developing technology. But, says Keel, they are the same ultraterrestrials playing the same tricks and leaving us, as always, hopelessly muddled. It is a practical joke on a cosmic scale.

This fascinating hypothesis does not provide all the answers. If ultraterrestrials are responsible, why do they need to show themselves at all? What is their mission? Conjecture along these lines could lead us into the wildest regions of science fiction, particularly as more and more cases of UFO sightings seem to contain an element of the paranormal.

Two police officers, Chief Deputy Sheriff William McCoy and Deputy Robert Goode, were driving along a highway in Brazoria County, Texas, in September 1965 discussing what to do about Goode's finger. It had been bitten by his son's pet alligator, and it was throbbing painfully. As they talked about waking a doctor at the end of their patrol, they spotted a large purple glow in the sky. It was a rectangular shaped object about 50 feet in height, and it turned and headed toward them. It was accompanied by a small blue light. Goode's window was down and he was waving his aching finger in the air. As the lights rushed toward them Goode felt a wave of heat on his hand and arm. The police officers, both frightened by the lights, accelerated and drove away. But McCoy looked back to see the UFOs rise upward, flare brilliantly, and disappear. When the two men got back to Damon, Texas, Goode realized his finger was no longer hurting him. He removed the bandage and discovered that not only was the wound no longer swollen or bleeding, but also that it was almost healed.

This cure for an alligator bite received much good-natured ridicule from the press, but healings associated with UFOs are not uncommon. Famous French researcher Aimé Michel reported the case of a doctor, who insisted on remaining anonymous. Dr. X had stepped on a mine during the Algerian War. As a result he suffered from partial paralysis of both limbs on the right side. He could no longer play the piano, at which he had been exceptionally gifted, and he walked with a pronounced limp. On a November night in 1968 he awoke at 3 a.m. to hear his young son crying. He got up to go to him. One of his legs was still swollen and painful from an accident while chopping wood three days before. After seeing to his son he noticed flashes of light outside, and on looking out saw two identical disks hovering in the sky. They had antenna from which flashes of light were being emitted at one-second intervals. As he watched the two disks merged into one and traveled toward him. A beam shone over him. He instinctively covered his eyes. There was a bang and the object dematerialized. Badly shaken, he went back to the bedroom to tell his wife about the strange object in the sky. He suddenly realized that the swelling and pain from his wood-chopping accident had vanished. Later the effects of his war wounds also disappeared. He now even plays the piano again.

Another indication of paranormal influence through apparent UFO activity came to light when a well-known Argentine attorney and his wife disappeared while driving south of Buenos Aires. Dr. Gerardo Vidal and his wife were traveling on the road from Chascomus to Maipu. When they failed to arrive at their destination friends and relatives searched the area but found no trace of them or their car. Two days later the doctor phoned from the Argentine consulate in Mexico City 4000 miles away. He reported that he had been driving from the suburbs of Chascomus when a dense "fog" appeared on the road in front of them. The next thing he remembered was sitting with his wife in their parked car on an unknown road. They both had a pain at the back of the neck and the sensation of having slept for hours. Their car looked as if it had been burned by a blow torch, and they had no recollection of what had happened in the intervening 48 hours or how they had been transported from Argentina to Mexico. Time-regression hypnosis may supply the answer one day, if the Vidals agree to such a test.

More than a quarter of a century after their modern debut, flying saucers are still a mystery. More confusion and doubt exists now than when Kenneth Arnold sighted nine shining disks over Mount Rainier in 1947. But Dr. Hynek and others are sure a continued scientific study is worth the effort. There is plenty of room for new investigators willing to dedicate their time to a specialized study of the various aspects of UFO phenomena.

"Although it will be a Herculean task to cull and refine existing UFO data," Dr. Hynek writes, "I feel a rich reward awaits a person or a group that assumes this task with dedication." He considers it possible that something might be found which represents a major scientific breakthrough. "It might call for reassignment and rearrangement of many of our established concepts of the physical world, far greater even than the rearrangements that were necessary when relativity and quantum mechanics demanded entrance into our formerly cozy picture of the world."

Travelers from Outer Space?

Below: Howard Menger sighted this not unattractive space creature near High Bridge, New Jersey, in June 1946. With long blond hair, standard human curves, blue eyes, dressed in a green semitranslucent pastel fabric, she was the height of an average Earth woman, and extremely graceful. Menger claimed he had sighted her 12 years earlier when he was 10 years old. She was friendly, and Menger later proposed to her and they were married.

More Travelers From Outer Space?

Above: near Desert Center, California on November 20, 1952. Sighted by George Adamski, UFO researcher. Was tracking a UFO when a man appeared. About 5 feet 6 inches tall, round-faced, slightly slanted eyes and high cheekbones. Had flowing shoulder-length hair which fell in waves and was of a sandy color. Wore a one-piece garment in a dark brown, which looked much like our ski suits. Greeted Adamski cordially, and subsequently visited him as a good friend.

Above: Caracas, Venezuela on November 28, 1954. Sighted by Gustavo Gonzales and José Ponce. Emerged from a luminous sphere. Four bristly, hairy, dwarflike creatures of amazing strength and impenetrable skin. One managed to push Gonzales 15 feet away with a single easy gesture. When Gonzales stabbed at it with his knife, the knife glanced off as if the skin were metal. All of them wore loin cloths. Their behavior was clearly hostile and aggressive.

Above right: Hopkinsville, Kentucky on August 22, 1955. Sighted by the Sutton family. Apparently emerged from a spaceship seen to land in a nearby field. There were several green ghostlike figures, about three feet tall, which seemed to glow from inside. Had round heads, great elephantine ears, and a mouth slit from ear to ear. Had no visible neck and the long arms ended in claws. Stood upright, but ran on all fours. The creatures clambered all over the roof of the Sutton house, and terrorized the family throughout the night.

Above left: near Francisco de Sales, Brazil on October 15, 1957. Sighted by Antonio Villas Boas, a farmer. Five creatures, the crew of an egg-shaped UFO. Nearly 6 feet tall, of normal strength. Small, light-colored eyes. Wore gray tight-fitting overalls which fastened to a stiff helmet with three silvery tubes emerging and entering the clothing. Had pineapple sized luminous red shield on the chest. Examined Villas Boas, took a blood sample, and used him, he said, for an interterrestrial breeding experiment with one of the crew.
Above right: Alto dos Cruzeiros, Brazil on October 26, 1962. Sighted by José Camilho Filho, a mechanic. Appeared sitting on a fallen banana tree, apparently associated with a UFO sighted nearby. Two three-foot tall creatures with brown shriveled skin, large head, white hair, and large slit eyes. Wore green trousers, blue shirts, and a luminous chest shield that flashed like a welding lamp. Took flight and ran, colliding with each other, when they saw Filho.

Above: Ashland, Nebraska on December 3, 1967. Sighted by Herbert Schirmer, a policeman. Came out of a football-shaped UFO. Four crew members, about five feet tall, with gray-white skin, flat noses, and a motionless slit mouth. Slightly slanted eyes that did not blink, but whose pupils widened and narrowed like a camera. Wore silvery-gray uniforms, boots, and gloves. Winged serpent emblem on the chest. Took Schirmer into the spacecraft, lectured him, and afterward induced amnesia.

Chapter 13
An Inhabited Universe?

In our vast and mysterious universe, are there other living beings with whom we could communicate? We have heard the stories of many who claim to have contacted spacemen: what is the strictly scientific opinion? How was life formed? Is it perhaps carried through space inside meteors and deposited on planets, where it develops if the conditions are favorable? Programs to investigate this hypothesis are now under way: what can they teach us? We are now broadcasting messages that tell of our existence and beaming them across the stars in the vast Milky Way. If there is an answer, how might it come? More importantly, what will it be?

President Jimmy Carter has seen a flying saucer. It happened in 1973 after he had made a late-evening speech to a Lions Club meeting in Thomaston, Georgia. As he left with a group of friends he saw the UFO, apparently hovering above a field.

"It was a very peculiar aberration, but about 20 people saw it," he told the *National Enquirer* newspaper during his presidential election campaign. "It was the darndest thing I've ever seen. It was big, it was very bright, it changed colors and it was about the size of the Moon. We watched it for 10 minutes, but none of us could figure out what it was. One thing's for sure, I'll never make fun of people who say they've seen unidentified flying objects in the sky!" Carter's 23-year-old son gave the newspaper further details about the UFO: "It had three lights—clustered together about the size of the Moon. My dad said they were changing color from red to green. It was positioned off to one side of the Moon."

It is now up to the American President, who is on record as saying "I am convinced UFO's exist because I have seen one," to take time from the burdens of his office and open the UFO files to public inspection—for millions of his countrymen believe that the American authorities know more about UFOs than they are saying.

Many other authorities adopt a secretive policy on UFOs, as can be seen from a statement made in July 1976 by another eminent UFO eye-witness, General Carlos Cavero, commanding

Opposite: the official agencies dismiss the continuing reports of UFO activity as unimportant and unsubstantiated, but there remain many who are convinced of their reality. Here two committed British UFO enthusiasts track through the countryside using special UFO-detecting saucers to pick up any trace of mysterious spacecraft.

The Mystery of Tunguska

Opposite: an artist's impression of the so far unexplained fireball that exploded in remote Siberian marshland on June 17, 1908. Unfortunately, it was nearly 20 years before any scientific investigation of the area was made. Even then, however, it was plain that a vast area had been devastated by the explosion—and most puzzling of all, there were no fragments of the fireball, which had been presumed to be a giant meteor. Whatever had burned across the Siberian sky had vanished, like a huge bomb. Even more mystifying was the more recent discovery that the rate of genetic change in the region shows a dramatic increase over the normal rate.

officer of the Spanish Air Force in the Canary Islands. While spending a few days near Sabada, a village in a remote part of eastern Spain, he saw a bright UFO traveling at a speed of 22,000 miles per hour. He calculated the speed from the time it took the UFO to travel the 12 miles to the next village—just two seconds. General Cavero said, after reporting his sighting to the Spanish Air Ministry in Madrid: "As an Air Force officer I hold, officially, the same opinions as the Air Ministry. But personally, I think these unidentified flying objects come from other planets."

The UFO seen by one of Spain's top serving officers, was also apparently seen by a well known local doctor, Francisco Padron, who had a close encounter with the object. "It was nearly dark," he said, "and I was driving along a lonely road. Then I saw a round sphere about 40 feet in diameter hovering above the road ahead of me. It was emitting a blueish light and was about 20 feet above the ground. As I approached, my radio cut out but my headlight kept working. I passed right underneath and saw, silhouetted inside through a type of porthole, two very tall figures dressed in bright red. Then it took off and vanished in the direction of Tenerife."

So, despite the official denials and the negative findings of investigations such as the Condon Committee, the flying saucer mystery deepens with an American president and a Spanish air force general among the eye-witnesses in the 1970s. The UFOs refuse to go away; so, again, what are they?

There are two modern mysteries to add to the already long list of mysteries for which there appears to be no official explanation. One relates to the cold wastes of Siberia and the other concerns a tribe living in sub-Saharan Africa. Either of these may in time prove to be a vital key in unlocking the secrets of UFO phenomena.

A giant ball of fire shot across Western China and Mongolia on the morning of June 30, 1908, and exploded with a deafening roar in the desolate marsh region of Tunguska in the heart of Siberia, 2200 miles east of Moscow. Seismographs as far away as Washington registered the event and a series of thunderclaps were heard 500 miles away. A "black rain" of debris fell over a wide area and trees were flattened over an area of 1500 square miles (whereas the Hiroshima atom bomb flattened only 18 square miles). It seemed to be a spectacular meteorite. But there is a mystery: A meteorite that could cause such devastation should have left a gigantic crater on impact, but no crater nor remains of the meteorite have been found. Because of Russia's internal problems, the Siberian explosion was not investigated for 19 years. When Leonid Kulik, a meteorite expert, led an expedition to the area in 1927 he found a charred forest with flattened trees—their roots pointing to the center of the blast. But strangely, the trees at the center were still upright. One explanation is that whatever it was that exploded did so high above Tunguska.

In 1976 the Russians announced that a new study of the mystery was to be carried out. This followed the discovery of very strange effects in the area. The Tunguska peat swamps receive their minerals only from the air and they grow at a regular rate, so it has been possible to examine the layer that

The Search for Evidence

formed at the time of the great explosion. Silicate particles have been found that are of a composition which not only differs from that found in ordinary meteorites, but has no equivalent in known earthly bodies. Genetic mutations have now been reported from the area and a mighty coniferous forest has blossomed in the once devasted region. The natural rate of genetic change in the region's flora has increased twelvefold, it is said, in which case some exciting discoveries could appear when the Russian expedition publishes its result.

What, then, caused the 1908 Tunguska explosion? It would be easy to dismiss it as just a supermeteorite, or a comet, or even an as-yet unknown cosmic object that strayed too close to the Earth and exploded in the atmosphere. Some scientists have suggested that it might have been an object made of antimatter that was annihilated by the Earth's matter in a fantastic explosion. Others believe it might have been a small exploding black hole. But there is one disturbing aspect of the Siberian mystery that does not fit any of these theories. From a study of eye-witness accounts of the great ball of fire, Dr Felix Ziegel of the Moscow Institute of Aviation has concluded that the object made a 375-mile arc in the air before crashing or exploding. "That is," says Dr Ziegel, "it carried out a maneuver." This has led some scientists to suggest that the Tunguska object was a spacecraft that got into difficulties and crashed or exploded. Perhaps it was the first probe sent out by beings in another solar system and the information it

Below and right: two photographs of the devastated trees in the Tunguska area. These pictures were taken some distance from the Stoney Tunguska River where the presumed meteor apparently struck: the explosion flattened trees for 1500 square miles, and wiped out two small villages.

sent back enabled them to build new spaceships—the present-day UFOs—that were able to carry out surveillance of our planet without difficulty.

Our astronomical knowledge has taught us that while some stars are still forming in clusters of gas and dust, there are also stars that are far older than our Sun. If these stars have planetary systems, then life could have evolved to our present level many thousands of years ago. If that is so, then Earth could have been visited over regular periods by space people. They may even have made contact with early civilizations and influenced them. That is the conclusion drawn by Robert Temple, a young American with a degree in Oriental Studies and Sanskrit, after a detailed study of the Dogon people who live in Mali, in the former French Sudan in Africa.

The Dogon people, says Temple in his book *The Sirius Mystery*, preserve a tradition of what seems to have been an extraterrestrial contact. The Dogon's most secret traditions all concern a "star" to which they give the same name as the tiniest seed known to them. The botanical name of the seed is *Digitaria*. Temple first came across this strange fact in an account written in 1950 by two eminent anthropologists, Marcel Griaule and Germaine Dieterlen of France, who had lived with the Dogon and won their confidence. The mystery was contained in· one passage: "The starting-point of creation is the star which revolves round Sirius and is actually named the '*Digitaria* star'; it is

Big Bang Over Nevada

On the night of April 18, 1962 at about 7:30 p.m. an explosion ripped across the Nevada sky. The flash was as bright as an atomic blast, and the noise shook the earth for miles. Was it an atom bomb test? A meteor? An enemy missile or aircraft? These logical questions were never answered by those who investigated the incident.

The first report of an odd UFO had come from Oneida, New York. Observers there saw a glowing red object moving west at a great altitude. It was too slow to be a missile, too high to be a plane. A meteor was ruled out because this object was tracked by radar, and meteors cannot be. As it moved west across the country, reports of it came in from the states of Kansas, Utah, Montana, New Mexico, Wyoming, Arizona, and California.

At some point the huge UFO landed near an electric power station in Eureka, Utah. Until it took off again, in its own time, the station was unable to operate at all.

The possibility that the explosion was from a nuclear test was denied by the Atomic Energy Commission. Its spokesmen said there was no atomic testing anywhere on the North American continent at that time.

Jet interceptors from the Air Defense Command pursued the UFO, but radar screens lost it about 70 miles northwest of Las Vegas. It was in that precise direction that the blast took place somewhere above the Mesquite Range.

Few people in the United States ever learned about this unusual event. Only the Las Vegas *Sun*, which was in the area of the explosion, carried the story. The news was otherwise suppressed by the Air Force.

The Sirius Riddle

Below: the first photograph of Sirius B, taken in 1970. It is the tiny dot to the lower right of the large star Sirius. Dr. Irving W. Lindenblad of the United States Naval Observatory finally succeeded in achieving this picture by an ingenious hexagonal lens that managed to compress the bright star's image so that the much smaller, dimmer star could "peek through." Yet, what scientists discovered mathematically only a century ago, and have only recently been able to see, a primitive sub-Saharan people, the Dogon, seem to have known about for hundreds of years. They have always drawn pictures of Sirius with a companion star and even more remarkably, they knew that it turned on its axis and that it followed a fixed orbit around its bright neighbor.

regarded by the Dogon as the smallest and heaviest of all the stars; it contains the germs of all things. Its movement on its own axis and around Sirius upholds all creation in space. We shall see that its orbit determines the calendar."

It is astonishing that a relatively primitive people should understand that stars revolve on their axis and follow orbits, but what is really incredible about the Dogon traditions is that the star to which they refer (known to astronomers as Sirius B) is totally invisible without the aid of a telescope, and was not discovered until the last century. It was only photographed in 1970 and it is so faint that even if it were not obliterated by the brilliant star Sirius it would still be invisible to the naked eye. The reason is that it is a white dwarf—the tiniest form of visible star in the Universe. At the time of its discovery in the middle of the 19th century it was unique, but a hundred or more have now been seen by astronomers and there must be many thousands more in our galaxy alone. Sirius B is a very dense star, the smallest and heaviest type (except for neutron stars, which are invisible), just as the Dogon say. According to their traditions it is made of a metal that is so heavy "that all earthly beings combined cannot lift it." That's a surprising concept but a correct one: A cubic foot of Sirius B matter would weigh 2000 tons. The Dogon also describe the elliptical orbit of the dark companion of Sirius and say that it takes 50 years to orbit the bright star, which has since been confirmed.

Sirius has always been an important star. It is the brightest in the sky and was the basis of the Egyptian calendar. So it would not be surprising for the Dogon or any other people to have myths about it. But what is uncanny is that the Dogon regard Sirius as relatively unimportant. It is described only as the focus of the elliptical orbit of the invisible *Digitaria*. What makes Temple's well-researched book so impressive is that much of the evidence supplied came from the two French anthropologists who were simply reporting the Dogon traditions without attempting to explain or understand them. It is clear that Griaule and Dieterlen had hardly any astronomical knowledge and did not realize that the Dogon traditions about the Sirius star system were incredibly accurate.

But why should a sub-Saharan tribe have sophisticated knowledge of exotic stars and their behavior? Temple offers evidence that the information the Dogon possess is more than 5000 years old and was possessed originally by the ancient Egyptians in pre-dynastic times before 3200 B.C. The Dogon, he says, are partially descended culturally and probably physically, too, from these people.

Although he attempts to give the facts without speculation, Temple admits that his findings "seem to reveal a contact in the distant past between our planet Earth and an advanced race of intelligent beings from another planetary system several light-years away in space." He puts that cultural contact at some time between 7000 and 10,000 years ago but admits that it may go back farther, and that civilization as we know it could even have been an importation from another planet in the first place.

"Nommo is the collective name for the great culture-hero and founder of civilization who came from the Sirius system to set up society on the Earth," says Temple, referring to the Dogon tradition. "The Nommos were amphibious creatures . . . more or less equivalent with the Sumerian and Babylonian tradition of Oannes [the demigod-leader of the amphibious Annedoti, creatures who brought civilization to Earth]."

If these interpretations are correct then at this very moment there is a largely ocean-covered planet in the region of Sirius just 8.6 light-years away, inhabited by creatures far in advance of us. They have visited Earth in the past and perhaps they are still keeping us under observation, possibly in conjunction with other superior beings with whom they enjoy intergalactic collaboration.

But let us for a moment put aside all the evidence of the Dogon, the Tunguska "thing" that exploded, and the thousands of unexplained UFO sightings, and look at the possibility of life elsewhere in the Universe from a different standpoint. We have to face the possibility that the Earth may be unique in supporting intelligent life. If that is so, then we have a tremendous responsibility to ensure the survival of the human race. If we are the first seed of life in the Universe then we may be destined to people other planets and spread life to other Solar Systems and even to other galaxies. But is that likely? The Universe that we observe is governed by cosmic laws that do not seem to encourage uniqueness. As astronomers discover a new cosmic object—a pulsar, a quasar, or a radio galaxy, for example—it is not long before they find others of the same type. So, since planets have formed around our Sun it seems certain that the same phenomenon will have occurred elsewhere in the Universe. It could be that every star has a planetary system, but even the nearest star is too far away for us to detect such orbiting, non-luminous bodies.

Dr R. N. Bracewell of the Radio Astronomy Institute of Stanford University has estimated that 10,000,000,000 planets in our galaxy alone could support life. Dr Melvin Calvin, of the Department of Chemistry, University of California at Berkeley, has said: "There are at least 100,000,000 planets in the visible Universe which were, or are, very much like the Earth. . . ." Life, according to these and other scientists, could be a common phenomenon throughout the Universe. But how does it form? The theory that has enjoyed support for many years suggests that while the Earth was still at a primitive stage of evolution, lightning flashes in the atmosphere would have supplied sufficient energy to break the chemical bonds of certain molecules on our planet causing a rearrangement of the atoms into amino acids and nucleotides from which living organisms are made. A similar event can be produced in a laboratory with electrical discharges.

In 1977 two astronomers suggested a new and exciting theory which, if confirmed, would seem that life is undoubtedly a widespread phenomenon. Fred Hoyle the British astronomer and Chandra Wickramasinghe, professor of applied mathematics and astronomy at University College, Cardiff, Wales, have suggested that life forms in space. Their theory is based on studies of meteorites and interstellar dust, particularly molecular clouds in which new stars are forming. They say: "We regard pre-stellar molecular clouds, such as are present in the Orion nebula, as the most natural 'cradles' of life. Processes occurring in such clouds

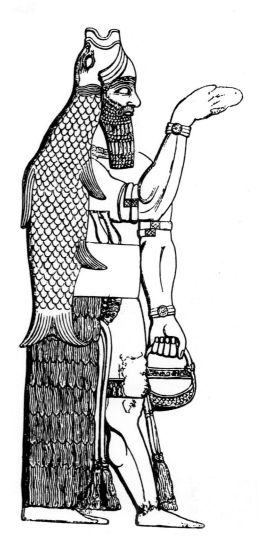

Above: the curious Babylonian demigod Oannes, a fishtailed amphibious being who, according to the Babylonian tradition, descended to found civilization on Earth. It was just such a being, or beings, that according to the Dogon people of Africa, arrived on Earth from the Sirius system to bring wisdom and culture to the inhabitants.

Meteors and Meteorites

Below: reports of meteors were dismissed once as superstitious nonsense—just as reports of UFOs are now. This 11th-century painting shows a lumpy asteroid falling past an astounded but fascinated observer.

lead to the commencement and dispersal of biological activity in the galaxy." They suggest that extremely complex molecules could be preserved in "grain clumps" that are carried through space inside meteorites and deposited on planets. At the moment, they point out, 100 tons of meteoritic material enters the Earth's atmosphere *every day*. Meteorites known as *carbonaceous chondrites* are believed to be the most primitive accumulations of solid material in the Solar System. They account for about half of all types of meteorite and are of particular interest because, say the two, "a few of the 20 biologically important amino acids have actually been identified in these meteorites."

In the past, theories that meteorites contain primitive life-forming substances have met with skepticism. Meteorites do not drop out of the sky and land at the feet of astronomers for instant study. Months or years may elapse before a meteorite is found and then it is difficult to know which substance it brought with it and which are the result of terrestrial contamination. One attempt to overcome this problem is a network of cameras arranged to photograph meteors and fireballs so that their trajectories can be established and their point of impact calculated. A European network, covering 425,000 square miles of Germany and Czechoslovakia has not made a single recovery in more than 13 years. The Canadian Research Council, Ottawa, runs a similar scheme which has been more successful. Known as the Meteorite Observation and Recovery Project (MORP), it has 16 camera stations that cover 308,000 square miles. It came into operation in 1971 and has succeeded in recovering a four pound 10 ounce meteorite that fell at Innisfree, Canada, on February 5, 1977. Once the information from the network was processed the search operation began and the meteorite was found within four hours (on February 17) just 545 yards from the predicted impact point. The recovery just 12 days after it fell greatly reduced the chances of contamination of the meteorite, a chondritic type. It was taken for study to laboratories around the world. No reports have yet been published, but it will be interesting to see whether they will lend support to the Hoyle-Wickramasinghe theory that "the transformation of inorganic matter into primitive biological systems is occurring more or less continually in the space between the stars." Another form of evidence would be the discovery of microfossils in the Innisfree meteorite. Such organized cell-like structures, microscopically similar to geological microfossils, were said to have been found by investigators in the early 1960s in carbonaceous chondrite meteorites, and though this interpretation was later disputed by other scientists, new supporting evidence has been offered that suggests that pollen-like spores *are* carried by meteorites.

So we might soon know for certain if meteorites and other cosmic bodies do scatter the seeds of life throughout the Universe. But whatever the origin of life, we know it has evolved on one planet, Earth, and the same may also be true of other planetary systems. Space travel is too slow a form of communication for beings on evolved planets unless they can travel at the speed of light. What seems more likely is that they would signal to each other over vast distances. They might even be trying to com-

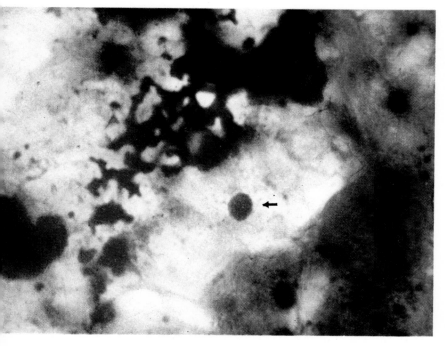

Left: a greatly magnified fragment of a meteorite. The arrow indicates a patterned area of rock like that produced by certain algae found on Earth. Any positive conclusion about the part that meteorites may play in distributing primitive life-forming substances must await recovery of a newly fallen meteorite before any contamination has taken place.

Below: the gigantic crater, nearly a mile wide, left near Winslow, Arizona, by a meteorite that fell some 30,000 years ago.

Messages From Outer Space

Below: Dr. Frank Drake, who in 1960 set up Project Ozma, the first attempt to listen to broadcasts from extraterrestrial sources. Two stars were studied at one radio frequency for a week. The results of this pioneer effort were negative.

Below: the huge 140-foot diameter radio telescope used for Project Ozma at the National Radio Astronomy Observatory, located in Greenbank, West Virginia.

municate with newly evolved civilizations, such as our own. But it was not until 1960 that Earth scientists began the first serious attempt to listen in to messages from other beings in the Universe. The man behind the attempt was Dr Frank Drake of the National Radio Astronomy Observatory at Green Bank, West Virginia. For 150 hours Dr Drake listened to the two nearest stars most like our Sun—Tau Ceti and Epsilon Eridani—with the observatory's 87 foot radio telescope. He found no trace of radio signals on the frequency he used. The search has continued, however, with larger radio telescopes and with other astronomers, including those in the USSR. Wavelengths monitored now range from radio to the ultraviolet, and the search has broadened from the nearest stars to distant galaxies.

For the first time, says Dr Carl Sagan, Professor of Astronomy and Space Sciences at Cornell University, man has the tools to make contact with civilizations on planets of other stars. "It is an astonishing fact," he writes in his book *The Cosmic Connection: An Extraterrestrial Perspective*, "that the great 1000-ft-diameter radio telescope of the National Astronomy and Ionosphere Center, run by Cornell University in Arecibo, Puerto Rico, would be able to communicate with an identical copy of itself anywhere in the Milky Way Galaxy."

Of course, there could be millions of civilizations capable of communicating, but if they all confine themselves to *listening* for messages then the Universe would be deceptively quiet. So the Earth's astronomers decided that as well as monitoring signals from other beings they should beam out a message about our tiny planet to anyone who may be listening.

Instead of aiming signals to one spot in the sky, a better way of announcing our scientific "maturity" is to set up transmitters that will radiate signals in the plane of our galaxy so that the vast majority of stars in the Milky Way will be swept, like a "lighthouse" beam, as the Earth rotates. This seems the most likely next step in the search for intelligent life in the Universe that is now advancing on a very broad front. Although no signals have been received in the first 17 years of searching this has surprised no one. Estimates suggest that man will have to extend his search to as far as 1000 light-years, surveying perhaps 100,000 stars, before receiving the first extraterrestrial message. So far, less than 1000 individual stars have been covered in the search.

If the entire hunt for intelligent life in the Universe fails to produce any sign of other civilizations will it mean that man is alone? It would be a disappointment, naturally, but it could be that other living beings are communicating in different ways using means that we have not yet discovered. Primitive men communicated over large distances by beating messages on a drum. Some peoples on Earth still use this method of communication, unaware that they are surrounded by radio messages and unable to tune in because they do not have radio receivers. Similarly, planets that are, for example, just 10,000 years ahead of us technologically may regard radio communication as a primitive method of making contact. If so, the Universe may be teeming with intelligent dialogues transmitted on wavelengths that we have not yet monitored or that have not yet been discovered by us.

Although the quest to find someone else in the Universe is important, the distances that separate the stars and galaxies make any form of transmitted conversation unlikely. If we are going to make contact with intelligent beings the chances are that it will come about as a face-to-face confrontation. And since it seems unlikely that advanced life exists in our Solar System and we are not going to reach other stars for many generations, the best hope of interplanetary communication will be in the form of visitors arriving on Earth. That brings us back to the UFO question and the attitude that governments should be taking. It does seem odd that while America and the Soviet Union are funding the radio search for intelligent life with vast amounts of

Below: the radio telescope at the National Astronomy and Ionosphere Center, run by Cornell University in Arecibo, Puerto Rico. The 600-ton triangular platform of the telescope—the world's largest and most sensitive—hovers over its 1000-foot diameter reflector bowl. In November 1974 it beamed out a radio message giving coded information about the Earth and its technical achievements at that date, aimed at a star cluster in the Milky Way.

Signaling into Outer Space

Opposite: the Arecibo message is in a binary code, which is decoded by breaking up the characters into 73 consecutive groups of 23 characters, and arranging the groups one under another. This results in this visual message, in which each 0 of binary code represents a white square and each 1 a black square. The translation on the right shows the message decoded.

Below and right: among the proposals for sophisticated listening posts is the Cyclops system, shown here in an artist's impression. It is a collection of hundreds of individual antennas linked to act in unison. The effective signal-collecting area of the system would be hundreds of times the area of any existing radio telescope, and has the advantage of using only existing technology: the necessary electronic and computer techniques are already well developed and in present use.

money, the possibility that extraterrestrials may be all around us seems to have been neglected apart from well-meaning but biased civilian investigations.

When the American Astronomical Society sent a questionnaire on UFOs to its 2611 members, 80 percent of the 1356 who responded said the phenomenon deserved scientific study. The results of the study, organized by Professor Peter Sturrock, a Stanford University, California, astrophysicist, were announced early in 1977. It means that at least 40 percent of the AAS total membership is in favor of a UFO investigation. What is more, 62 of the astronomers who replied claimed to have seen a UFO. Five of these witnessed unidentified objects through telescopes and a further three made sightings with binoculars. In seven cases there were even photographs of UFOs, though Professor Sturrock believes he can explain two of them in normal terms.

Not all the astronomers supported the investigation. It was regarded as unnecessary by 20 percent and some members were very outspoken in their criticism. "I object to being quizzed about this obvious nonsense," said one, while another insisted that "there is no pattern to UFO reports except that they predominantly come from unreliable observers." Professor Sturrock is a staunch supporter of a renewed investigation of UFOs and is a strong critic of the Condon Committee report, which brought the USAF's "Project Blue Book" to an end.

Meanwhile, as the UFO mystery persists here on Earth, our planet has launched its own "UFO" into interstellar space in the form of Pioneer 10. It left Earth in 1972 on a trajectory that took it across the orbit of Mars, through the asteroid belt and on to Jupiter, which deflected the path of Pioneer 10 and swung it past the orbit of Uranus and on out of our Solar System. It will continue to travel for 2,000,000 years before approaching the star Aldebaran, and for 10,000,000,000 years before entering the planetary system of another star—unless it is intercepted earlier by space travelers. If it is found, the extraterrestrial intelligences who investigate this "unidentified flying object" in space, or in

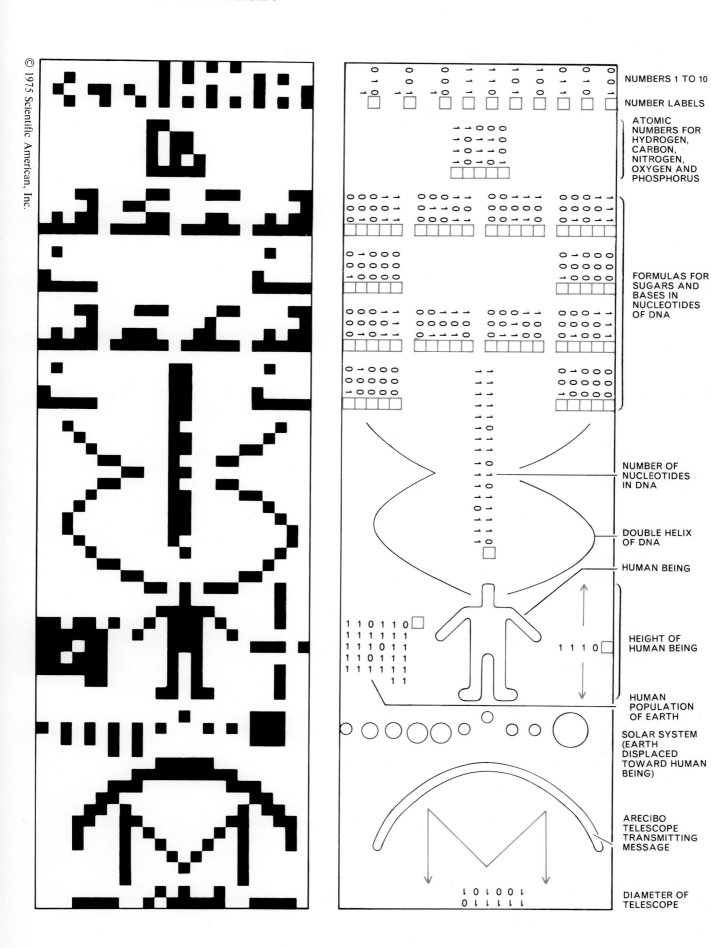

Earth's Message to the Universe

their sky, should be able to discover its origin very quickly. The American space scientists placed a gold-anodized aluminum plaque on Pioneer 10 with an etching showing the Solar System in a coded form with clues which, though they would mystify most Earth people, are designed to tell a superior race exactly where the spacecraft originated. The plaque included line drawings of a naked man and woman to show the beings responsible for launching the craft.

Dr Carl Sagan was the man who conceived the message from Earth and his artist wife, Linda, drew the terrestrials on the golden greeting card. He described his special "communications" project as "very much like a shipwrecked sailor casting a bottled message into the ocean—but the ocean of space is much vaster than any ocean on Earth." Nevertheless, messages in bottles have been washed ashore in time to save shipwrecked mariners, so there is a remote chance that Pioneer 10 will meet up with intelligent beings on a distant shore. The likelihood is, however, that when that happens man will have vanished from his "island" in space.

While we wait for a reply to our space messages, however, scientists are continuing to develop instruments and space exploration techniques that will help to solve the many mysteries of the Universe, some of which have puzzled us for centuries while others have only just begun to tantalize us. One fascinating new type of object that is high on the list for investigation is the X-ray burster. These were first detected after an analysis of records from a Dutch satellite in December 1975. Then astronomers from the Massachussetts Institute of Technology (MIT) examined records from the SAS-3 satellite launched in May of that year and discovered the same bursts. These X-ray sources suddenly flare up and die down in seconds, making them difficult to detect. So far 30 such bursters have been found. Dr Walter H. G. Lewin of the MIT, explained in 1977: "There is something in nature that can produce energy greater than a million suns and can turn it on in just one second." These explosions of "mind-boggling" power are all taking place in our own galaxy. Astronomers are now hoping to find objects, such as stars, at the points where the astonishing X-ray bursts are occurring. One of the sources, MXB 1730-335, is bursting at an astonishing rate: The time between each flare up varies from about six seconds for small bursts to between five and ten minutes for massive bursts.

What are the X-ray bursters? And what about the other questions we posed such as, What are quasars? How was the Earth formed? Are there such things as black holes? Can anything go faster than light? Does life form in outer space? Can a comet become a planet? Will the Universe expand for ever? Will man ever know all the answers or will there always be new mysteries to challenge his ingenuity and understanding? The next few years may provide some surprising answers.

The mystery of unidentified flying objects is a different matter, for even if every sighting of a "flying saucer" could be proved to be no more than misidentification, hallucination, or deliberate fabrication, the next one—arriving from somewhere deep in our mysterious Universe—could be for real.

Above: Dr. Carl Sagan, who is director of planetary studies at Cornell University. He is one of the foremost spokesmen of exobiology—a new field of study dealing with the possibility of extraterrestrial life and the means for detecting it.

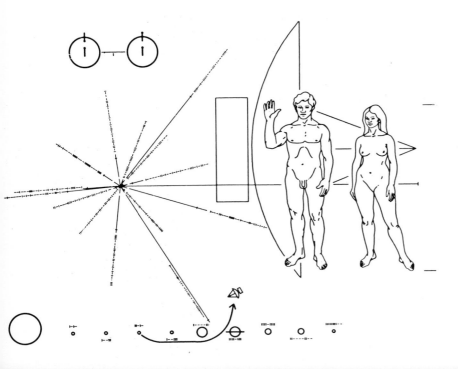

Left: the drawing on the anodized gold plaque that was put aboard Pioneer 10, the first spacecraft launched to travel beyond our Solar System. The coded message, designed to be intelligible to any beings who are sufficiently advanced scientifically to be able to intercept and recover Pioneer 10, shows our Solar System, indicating the planned trajectory of the craft into limitless space.

Below: Pioneer 10, launched in 1972. It passed Mars and Jupiter, which deflected its path and swung it out of our Solar System, set to travel for millions of years before approaching the nearest star.

Index

References to illustrations are shown in italic.

Picture Credits